FUNDAMENTALS
OF
STATISTICS

FUNDAMENTALS
OF
STATISTICS

H. MULHOLLAND, M.Sc., F.I.M.A.
Liverpool Polytechnic

C. R. JONES, M.Sc., F.S.S.
Liverpool Polytechnic

LONDON
BUTTERWORTHS

THE BUTTERWORTH GROUP

ENGLAND: BUTTERWORTH & CO. (PUBLISHERS) LTD.
LONDON: 88 Kingsway, WC2B 6AB

AUSTRALIA: BUTTERWORTH & CO. (AUSTRALIA) LTD.
SYDNEY: 20 Loftus Street
MELBOURNE: 343 Little Collins Street
BRISBANE: 240 Queen Street

CANADA: BUTTERWORTH & CO. (CANADA) LTD.
TORONTO: 14 Curity Avenue, 374

NEW ZEALAND: BUTTERWORTH & CO. (NEW ZEALAND) LTD.
WELLINGTON: 49/51 Ballance Street
AUCKLAND: 35 High Street

SOUTH AFRICA: BUTTERWORTH & CO. (SOUTH AFRICA) (PTY.) LTD.
DURBAN: 33/35 Beach Grove

First Published 1968
Second Impression 1969
Third Impression 1971

Suggested U.D.C. number: 519·2

ISBN 0 408 49200 7

Printed in Great Britain by
Page Bros. (Norwich) Ltd.

PREFACE

This book is intended primarily for the use of students studying for the General Certificate of Education at Advanced Level, for Higher National Certificates in Mechanical or Electrical Engineering, Mathematics, Chemistry, Biology or Pharmacy, or for a University or Technical College (C.N.A.A.) degree in any of the subjects mentioned. It could also be used as a good basis for any introductory statistics course.

The early chapters introduce the ideas of a sample and of randomness and the reader is urged to carry out his own practical experiments to obtain his data and verify the theory. Probability, including the idea of $\mathscr{E}(x)$ and its relation to the mean, and the related Binomial and Poisson distributions are dealt with thoroughly before continuous distributions are considered. Sample mean as an estimate of central tendency and its similarity to, but distinction from $\mathscr{E}(x)$, is thoroughly covered. The introduction of variance has been left until Chapter 6, where it is associated with mathematical expectation via $\sigma^2 = \mathscr{E}[(x - \mu)^2]$ in order to emphasize its greater importance over other measures of dispersion. Truly standard deviation only has worthwhile importance for normal and quasi-normal distributions and its use with Binomial and Poisson distributions is deferred until after discussion of its use with continuous distributions.

The distinctions between $\bar{x}$ and μ and between s^2 and σ^2 are discussed in Chapter 15 as also is the important theorem on the sampling distribution of sums, differences and products of two variates and the deduction of the standard error of the mean. This chapter could profitably be read earlier by the more mathematically able student.

Tests of significance are discussed using only tests based on the normal distribution, but such tests are covered thoroughly for the mean and variance. The most simple test is first considered and correspondingly more difficult ones are then introduced. Confidence intervals are dealt with at the same time as significance testing. Chapters on bivariate distributions, quality control and time series are included to give an idea of some of the more practical applications of statistics.

A collection of tables is included at the end of the book. They include, among others, tables of Binomial and Poisson probabilities.

The book contains many examples and we should like to express our thanks to the Northern Universities Joint Matriculation Board (J.M.B.) and Liverpool University (Liv.) for granting us permission to quote questions from their examination papers. The above abbreviations have been used to indicate the source of such questions. We our thanks to the Cambridge University Press for permission to reproduce their tables.

Finally we wish to thank the publishers for the care and trouble they have taken over the general presentation of the text.

C.R.J.
H.M.

CONTENTS

1

INTRODUCTION

THE word statistics is used to describe the collection, reliability, organization, representation, analysis and interpretation of data and not just the collection of numerical results.

Organization and representation of data apply to the ordering and presentation of the collected measurements in the form of tables or graphs suitable for further analysis. Analysis of the data is carried out in several ways, some of which are considered in this book. Interpretation is applied to the results of the analysis and includes the drawing of conclusions. Since these conclusions must be reliable, the data collected must be representative of the whole mass of possible data to which the conclusions are to be applied.

Statistics is, above all, a tool to be used in many sciences. Understanding the processes of Statistics alone is of little practical value. To be of value statistics must be applied and a knowledge of the particular science in which the technique is to be used is essential.

It is also important to realize that the application of a logical process of mathematics, of which statistics is a branch, to a collection of data, does not of itself guarantee that the conclusions drawn are valid if the original data are incorrect or unreliable. For example, suppose that it is desired to find the cost per year of a weekly return journey from Edinburgh to London. Any data gathered as the result of one trip, however carefully the resulting expenses were recorded, would not, when multiplied by 52, give the yearly cost. Account must be taken of variation in prices or change in the mode of travel. Again, to draw conclusions as to the outcome of a Parliamentary Election from a sample of voters obtained by telephoning 100 people would be wrong. Not only is the sample too small, but it is also not representative of the population, it is biased towards one section. In this book we shall consider the science of statistics, but it is the *user* of statistics who must first be satisfied that he has reliable data.

Statistics was originally applied to the data gathered by the State for the purposes of government. Interest in statistics was stimulated, at the beginning of the nineteenth century, because of the need to raise taxes to pay for the Napoleonic Wars. Large amounts of information were required in order to assess the future revenue of the country. The more mathematical side of the subject was based on work

1

done earlier by a number of eminent mathematicians such as Pascal (1623–1662), James Bernouilli (1654–1705), and probably the greatest of all the earlier writers on statistical subjects Gauss (1777–1855).

Since the beginning of this century the subject has received much wider attention as its usefulness, first to agriculture and then to industry, has been realized. Nowadays, most large industrial and commercial firms use some statistical techniques and in the training of scientists, engineers, economists and accountants a course in statistics is generally included.

It has already been emphasized that the collection of data to which the statistical processes are applied must be reliable. It must also be remembered that the use of statistical procedures does not, of itself, guard against faulty reasoning or incorrect interpretation of results. The misuse of statistics to prove that a desired conclusion is correct is not unknown.

2

THE COLLECTION, ORGANIZATION AND REPRESENTATION OF NUMERICAL DATA

2.1. THE COLLECTION OF DATA

As mentioned in the introduction, the study of statistics was greatly stimulated by the need for information concerning the population of a country. This information is necessary for forward planning of civil and military needs and also, perhaps the greatest spur, revenue forecasting. A study of a copy of the Annual Abstract of Statistics published by H.M. Stationery Office will indicate the great range of information which is now obtained and presented by different government departments. In a very much smaller unit such as a school, information is available concerning such things as ages, weights and heights of pupils, the number of meals served each day, and examination marks.

The information (data) is normally in a numerical form. It could be the number of people or objects with a given characteristic, e.g. the number of people in the school who have eyes coloured green, blue, brown, etc., or a measure of a characteristic, e.g. the heights of the people in the school.

Exercises 2a

1. On any one page of a book of poems, count the number of words with one letter, with two letters, with three letters, etc. Repeat the count for one page of a book of prose and also for one page of a book of plays. How can the results be depicted? (Keep a copy of your results.)

2. Measure the heights and weights of all pupils in the class. (Keep a copy of your results.)

3. Ask each member of the class to bring either an orange, an apple, a pear or a banana. (*a*) Count how many of each. (*b*) Weigh each fruit and record the weight.

With the introduction into industry of mass production methods continuous collection of data is necessary in order to keep a check on the quality of the output. In large firms there are masses of data which have to be collected regarding the number of employees, their wages, amounts of raw materials used, the output of the different products etc. In order to proceed further the meanings of several important terms must first be discussed.

Population

We use the word *population* to describe all possible measurements of the particular characteristic under consideration. A population can be finite (small or large) or infinite (in the sense that it is practically impossible to count its size). For example, the number of pupils in a class (small), the yearly output of an automatic machine (large), the number of grains of sand in the world (practically infinite).

Variate

By *variate* is meant the measurement of the characteristic which is being considered. A variate can be either *discrete* or *continuous* and is usually denoted by x. A discrete variate is one which is limited to certain possible values only and involves counting. For example, examination marks or the number of peas in a pod. A continuous variate can take any value in a given range and involves measurement. For example, in the case of barometric and temperature readings or percentage of impurities in raw materials, the variates can alter by amounts as small as we like and the variate is said to be continuous.

Since all measurements can, in practice, only be measured to a certain degree of accuracy they are to all intents and purposes discrete. In this book discrete data is dealt with first.

Frequency

In any population two or more members may have the same value. For example, the age (to the nearest day) of several members of a school may be the same. The number of members with the same value is known as the *frequency* and is generally denoted by f.

2.2. THE CLASSIFICATION OF DATA

Table 2.1* gives the heights, weights and colour of hair of a class of 35 pupils.

A large amount of data presented with no attempt at orderliness as in *Table 2.1* is of little value. The data can be put into a much more readable form if we give the frequency of occurrence of the different values of the variate. The heights of pupils in *Table 2.1* can be classified as shown in *Table 2.2*. Note that the fifth marks are used to cross out the previous four. This helps the counting of the number of marks for each height.

Make a similar count for the *weights* given in *Table 2.1* and hence complete *Table 2.3*.

Having filled in *Table 2.3* it can be seen that because of the number of values of the variate involved we do not obtain a very satisfactory

---

* See if you can collect this data for the members of your own class.

Table 2.1

Pupil	Height (in)	Weight (lb)	Colour of of hair	Pupil	Height (in)	Weight (lb)	Colour of hair
1	53	70	Red	19	55	81	Light brown
2	51	65	Light brown	20	49	63	Black
3	56	77	Light brown	21	55	68	Dark brown
4	54	69	Dark brown	22	53	66	Dark brown
5	52	66	Dark brown	23	56	80	Black
6	57	71	Dark brown	24	59	69	Dark brown
7	55	73	Golden	25	49	75	Light brown
8	48	61	Light brown	26	58	75	Auburn
9	50	64	Dark brown	27	54	72	Light brown
10	54	70	Auburn	28	58	78	Light brown
11	53	61	Black	29	57	74	Black
12	53	67	Light brown	30	52	67	Auburn
13	52	72	Dark brown	31	52	74	Light brown
14	55	73	Dark brown	32	51	68	Dark brown
15	54	76	Light brown	33	55	70	Dark brown
16	57	74	Light brown	34	54	77	Black
17	56	76	Light brown	35	54	80	Light brown
18	53	71	Dark brown				

Table 2.2

Height (in) x	Frequency f	
48	/	1
49	//	2
50	/	1
51	//	2
52	////	4
53	////	5
54	//// /	6
55	////	5
56	///	3
57	///	3
58	//	2
59	/	1
	Total	35

picture of the distribution of weight. It is better to group the values of the variate into *classes*. A class contains several values of the variate; in *Table 2.3* if we group the data into classes 61–63, 64–66, . . . 79–81, we obtain *Table 2.4*.

Table 2.3

Weight to nearest (lb)	61	62	63	64	65	66	67	68	69	70	71
Frequency											

Weight to nearest (lb)	72	73	74	75	76	77	78	79	80	81
Frequency										

Table 2.4

Weight to nearest lb x	Frequency f
61–63	3
64–66	4
67–69	6
70–72	7
73–75	7
76–78	5
79–81	3

Table 2.4 gives a better indication of the distribution of weight than *Table 2.3*. The number of classes will depend on the total frequency, but a general rule would be to have between 6 and 20 classes.

The number of values of the variable included in each class need not be the same throughout the distribution. For example, in considering the wages of the members of a small manufacturing company which employs part-time workers, the first class could be fairly wide to include all the part-time workers, and subsequent classes smaller to give more detailed information about the full-time staff.

Exercises 2b
1. The *Table 2.5* gives the weights to the nearest 0·1 g of 75 pieces of scrap metal from an engineering process. The weights range from 0·1 to 3·6 g. Make a table showing the frequency with which each weight occurs. Classify the data into 12 classes. Does the distribution appear to have any particular shape? Keep a copy of your results for use later.

2. *Table 2.6* gives the lengths to the nearest 0·01 in of 50 screws. Classify the data into (*a*) 16, (*b*) 8 classes. Which classification gives the better picture of the data? Keep a copy of your results for use later.

Table 2.5

1·7	2·0	0·8	1·7	1·3	0·3	2·1	1·4	1·5	3·0	0·8	1·8	1·4	1·7	1·4
3·5	1·3	1·9	1·2	2·2	3·3	2·6	1·2	2·5	1·6	2·0	1·2	1·3	1·4	1·1
1·3	1·1	2·3	0·1	1·1	1·7	2·7	1·2	1·4	2·9	0·6	2·8	0·7	2·0	1·1
0·8	1·0	1·2	1·0	1·5	1·6	1·0	2·8	0·7	0·5	1·5	3·0	1·6	1·5	1·6
2·6	1·5	0·5	1·6	3·1	1·9	1·9	0·9	2·4	3·4	1·8	1·5	1·3	0·4	2·5

Table 2.6

3·44	3·15	3·37	3·38	3·35	3·21	3·65	3·47	3·42	3·28	3·09
3·54	3·30	3·00	3·68	3·34	3·46	3·10	3·70	3·59	3·26	3·43
3·53	3·12	3·29	3·32	3·25	3·40	3·48	3·17	3·56	3·07	3·50
3·45	3·57	3·05	3·49	3·61	3·66	3·41	3·31	3·52	3·19	3·23
3·79	3·62	3·39	3·73	3·27	3·63					

Although putting the data into classes gives a better overall picture of the data, it destroys some of the original detail. For example, from *Table 2.4* we know that there are four weights which are either 64, 65 or 66 lb, but we cannot tell without referring back to *Table 2.3* what the actual values are. Thus when analysing data arranged in classes, as in *Table 2.4*, it is necessary to ascribe some value which will apply to all members of a given class. The most usual value is the middle value of the variate in the class (the *mid-interval* value). Thus for further mathematical analysis *Table 2.4* would be treated as if it were as given in *Table 2.7*.

Table 2.7

Mid-interval value of x	f
62	3
65	4
68	6
71	7
74	7
77	5
80	3
	35

B

Example 2.1. Table 2.8 gives the lengths to the nearest one thousandth of an inch of 100 screws. Put the data into classes (use 10 classes) indicating the value ascribed to the various members of each class for the purpose of mathematical analysis.

Table 2.8

3·293	2·617	3·095	3·087	2·803	2·700	3·134	3·056	3·061	3·152
3·149	2·962	2·953	2·612	2·937	2·939	3·126	2·783	3·362	3·153
2·992	2·970	2·761	2·804	2·942	2·710	3·122	2·614	2·771	3·137
3·075	2·926	3·197	2·913	3·189	3·045	3·179	3·148	3·199	3·472
3·387	2·619	3·083	2·755	2·513	3·184	2·907	2·904	3·041	3·168
3·282	3·062	3·267	2·811	2·855	2·847	3·118	2·800	3·166	3·349
2·981	2·617	2·831	2·745	3·213	2·713	3·221	2·613	3·020	3·259
3·335	3·105	3·317	2·800	2·863	2·732	3·015	3·231	3·245	3·026
3·112	2·891	2·827	2·872	3·214	2·900	2·900	3·206	3·034	3·459
2·516	3·102	3·275	2·884	3·007	3·008	3·421	3·205	2·618	3·309

Since the least value is 2·513 and the greatest 3·472, classes can be chosen as given in *Table 2.9*. These are, of course, not the only classes which could be taken. For the chosen classes we define the *class boundaries* as 2·4995–2·5995, 2·5995–2·6995, ... and the *class limits* as 2·500–2·599, 2·600–2·699, ...

The *class width* is defined as the difference between the smallest value of the variate in the interval and the smallest value of the variate in the next interval, or the difference between the upper and lower bounds. Equal class widths are not necessarily taken throughout a population.

Example 2.2. The frequency table (2.10) gives the weekly wages to the nearest shilling of all the 80 employees of the XYZ Company. Find as accurately as possible the total weekly wages bill for the employees.

The actual weekly wage of each employee cannot be found from the table. We can, however, make an estimate of the weekly wages bill by assuming that the members of each class earn the middle value of that class (see *Table 2.11*).

The approximate weekly wages bill is thus £1150.

The arrangement of data in a frequency table goes a long way to clarifying the original mass of values. However, we can obtain an even clearer picture if we represent the frequency table pictorially. Some of the more common types of pictorial representation are discussed in the next section.

Table 2.9

Length of screw to nearest 0·001 in x	f	Value ascribed for the purpose of further analysis
2·500–2·599	2	2·5495
2·600–2·699	7	2·6495
2·700–2·799	9	2·7495
2·800–2·899	13	2·8495
2·900–2·999	14	2·9495
3·000–3·099	15	3·0495
3·100–3·199	19	3·1495
3·200–3·299	12	3·2495
3·300–3·399	6	3·3495
3·400–3·499	3	3·4495
	100	

Table 2.10

Wages to the nearest shilling s	Frequency f
160–199	18
200–239	25
240–299	17
300–399	10
400–599	6
600–999	4
	80

Table 2.11

f (1)	Assumed weekly wages (shillings) (2)	Total wages (shillings) col. (1) × col. (2)
18	179·5	3231
25	219·5	5487·5
17	269·5	4581·5
10	349·5	3495
6	499·5	2997
4	799·5	3198
		22,991

2.3. GRAPHICAL REPRESENTATION OF DATA

Bar Charts

To construct a bar chart to illustrate a frequency distribution, parallel bars of equal width, whose *lengths* are proportional to the frequency are drawn horizontally, starting from the same vertical line. The widths between the bars being one half the width of a bar.

Example 2.3. From the data of *Table 2.1* construct a bar chart showing the number of pupils with each colour of hair.

From *Table 2.1* we obtain *Table 2.12* which is illustrated as a bar chart in *Figure 2.1*.

Table 2.12

Colour of hair	Light brown	Dark brown	Black	Auburn	Golden	Red
Frequency	13	12	5	3	1	1

This type of chart is normally used to compare data which refer to different types of categories.

The Pie Chart

A pie chart is only used when the data contains *all* subdivisions of the subject being considered. Each division of the subject is expressed as a percentage (P per cent) of the whole and is represented by a sector of a circle. The angle of the sector is P per cent of 360°. A pie chart

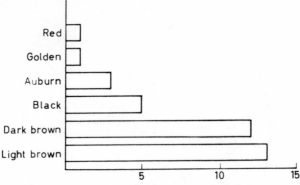

Figure 2.1. To illustrate the data of Table 2.12

should not consist of too many segments (less than eight is suggested) and each one should be clearly labelled.

Example 2.4. Construct a pie chart to illustrate the following data. The total male population over 21 years old in a certain town was 50,000. Of these, 43,400 were fully employed, 2000 were employed part-time, 4100 were retired or had private means, 500 were unemployed.

To calculate the angles of the sectors allotted to each category we proceed as follows:

Table 2.13

	Frequency (100s)	Fraction of the total frequency	Angle of sector
Full-time employed	434	$\dfrac{434}{500}$	$\dfrac{434}{500} \times 360 = 312 \cdot 5°$
Part-time employed	20	$\dfrac{20}{500}$	$\dfrac{20}{500} \times 360 = 14 \cdot 4°$
Retired or Private means	41	$\dfrac{41}{500}$	$\dfrac{41}{500} \times 360 = 29 \cdot 5°$
Unemployed	5	$\dfrac{5}{500}$	$\dfrac{5}{500} \times 360 = 3 \cdot 6°$
	500		

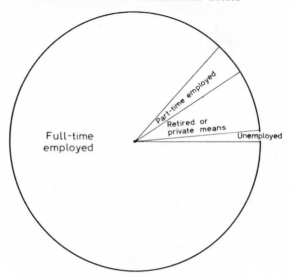

Figure 2.2. To illustrate the data of Table 2.13

Isotype Representation (Ideograph)

Pictures are often used for popular presentation of information but they can easily be used to misrepresent the data. Any differences in frequency should be indicated by a corresponding difference in the *number* of objects drawn—*isotype representation*.

Diagrams in which the *size* of the object drawn indicates its frequency are liable to misinterpretation.

Example 2.5. Construct an isotype representation (ideograph) of the following data:

Table 2.14

Year	1800	1840	1880	1920	1960
Population (1000s)	2	5·5	6·6	10·8	12·5

A misleading diagram is shown in *Figure 2.4*, where the *height* of the drawing of a man is proportional to the frequency. The diagrams suggest a volumetric representation, hence the relative frequencies have been exaggerated. The correct diagram is shown in *Figure 2.3*.

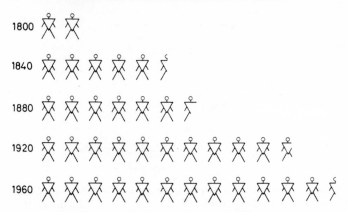

Figure 2.3. To illustrate the data of Table 2.14

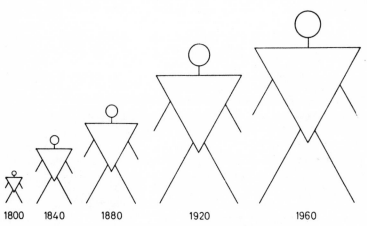

Figure 2.4. A misleading illustration of the data of Table 2.14

Histograms

A *histogram* is made by constructing rectangles on a continuous base, as shown in *Figure 2.5*, whose *areas* are proportional to the frequencies in each class. The bases of the rectangles being equal to the class widths. It is to be preferred when the variate is continuous or when the data has been grouped (classified).

Example 2.6. The life in hours of 100 television tubes is given in *Table 2.15*. Construct a diagram to illustrate the data.

Table 2.15

Life (h)	Frequency
200–399	6
400–499	8
500–599	12
600–649	27
650–699	25
700–799	14
800–1199	8
Total	**100**

This is an example of a continuous variate and a histogram is the best form of illustration. It must be again stressed that the *areas* of the columns are proportional to the frequencies. The calculation of the heights of the columns can be set out as follows in *Table 2.16*.

Table 2.16

Life (h) (i)	Frequency (ii)	Class widths (iii)	Height of column (iv) = (ii)/(iii)
200–399	6	200	$\dfrac{6}{200} = 0{\cdot}03$
400–499	8	100	$\dfrac{8}{100} = 0{\cdot}08$
500–599	12	100	$\dfrac{12}{100} = 0{\cdot}12$
600–649	27	50	$\dfrac{27}{50} = 0{\cdot}54$
650–699	25	50	$\dfrac{25}{50} = 0{\cdot}50$
700–799	14	100	$\dfrac{14}{100} = 0{\cdot}14$
800–1199	8	400	$\dfrac{8}{400} = 0{\cdot}02$
Total	**100**		

Note: in the above table the class widths are 199·5–399·5, 399·5–499·5, etc.

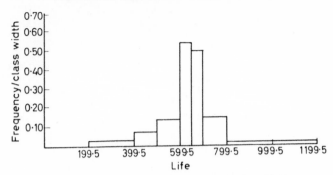

Figure 2.5. To illustrate the data of Tables 2.15 and 2.16

Example 2.7. The frequency distribution of marks of 120 boys in an examination were as given in *Table 2.17*. Construct a histogram to illustrate the data.

Table 2.17

Mark	Frequency
0–4	0
5–9	4
10–14	6
15–19	10
20–24	14
25–29	24
30–34	28
35–39	19
40–44	9
45–49	6
	120

This is a case of a grouped discrete variate and there are no marks between 4 and 5, 9 and 10 . . . so we make our class boundaries -0.5–4.5, 4.5–9.5, 9.5–14.5, . . . In *this case* the bases of the rectangles forming the histogram are all equal, therefore it happens that the heights will be proportional to the frequencies, *but* it must be remembered that it is the area which is proportional to the frequency (see *Figure 2.6*).

The Frequency Polygon (for Equal Class Intervals)

To construct a *frequency polygon* we mark off the class intervals on a suitable horizontal scale. At the centre of each interval we measure

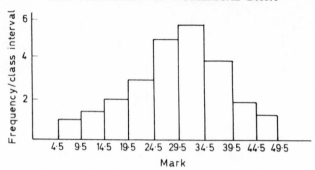

Figure 2.6. To illustrate the data of Table 2.17

ordinates whose heights are proportional to the frequency. The tops of the ordinates are joined to form the frequency polygon.
Note: This is equivalent to joining the mid-points of the tops of the histogram.

This type of representation is more often used with *time series*, that is when the distribution of the variate is given as a function of time.

Example 2.8. Illustrate the following data (*Table 2.18*) by constructing a frequency polygon.

Table 2.18

Years	Increase (1000's)
1800–1820	5·0
up to 1840	7·7
up to 1860	9·1
up to 1880	13·3
up to 1900	19·0
up to 1920	20·9
up to 1940	21·3
up to 1960	21·7

A broken-line graph can give some idea of the rate of change in a given interval. In the above example, with an interval of 20 years it gives only the average rate of change. The increase in any *one* year could bear no relation to the average shown in *Figure 2.7*.

Cumulative Frequency Distributions (Ogives)

The *cumulative frequency* corresponding to any value of the variate

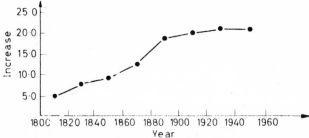

Figure 2.7. To illustrate the data of Table 2.18

x is the number of members of the population with a value less than or equal to x. In the case of a grouped distribution the cumulative frequencies are calculated for the upper boundary of each class. The values are plotted against the upper class boundaries and the points joined by straight lines.

Example 2.9. Construct the cumulative frequency distributions for the data of *Tables 2.15* and *2.17*. Draw the corresponding ogives.

The calculations are shown in *Tables 2.19, 2.20* and the ogives in *Figures 2.8, 2.9*.

Table 2.19

Life (h)	Frequency	Cumulative frequency
200–399	6	6
400–499	8	14
500–599	12	26
600–649	27	53
650–699	25	78
700–799	14	92
800–1199	8	100

Exercises 2c

1. Draw histograms to illustrate the frequency distributions obtained as the answers to Exercises 2(a) questions 1 and 2.

2. Draw histograms to illustrate the data of *Tables 2.9* and *2.10*.

3. Construct pie charts and bar charts to illustrate the data given in *Table 2.21*.

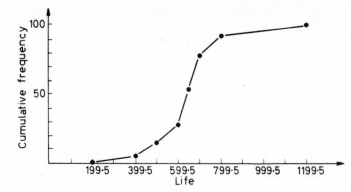

Figure 2.8. To illustrate the data of Table 2.19

4. Which kinds of pictorial illustration could be used to depict the following numerical data?

(*a*) The population (in millions) of Great Britain for the last 10 years?

(*b*) The total bank deposits made in the Greater London area for each five-year period from 1930 to 1960.

(*c*) The proportions of one's life spent in various activities: sleeping, playing, working, eating, etc.

(*d*) Seasonal rainfall (in inches at Anytown) for a 50-year period.

Table 2.20

Mark	Frequency	Cumulative frequency
0–4	0	0
5–9	4	4
10–14	6	10
15–19	10	20
20–24	14	34
25–29	24	58
30–34	28	86
35–39	19	105
40–44	9	114
45–49	6	120

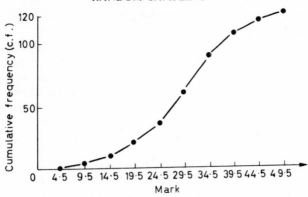

Figure 2.9. To illustrate the data of Table 2.20

Table 2.21

(a)

Continent	Area (millions of square miles)
Africa	11·7
Asia	17·0
America	16·3
Europe	3·9
Antarctica	5·3
Australia	3·0

(b)

Religion	Adherents (millions)
Buddhism	268
Confucianism and Taoism	300
Easter Orthodox	220
Hinduism	303
Muslims	402
Protestants	210
Roman Catholic	423
Others	146

2.4. RANDOM SAMPLING

Only in exceptional circumstances is it possible to consider every member of a population. In most cases only a sample of the population can be considered and the results obtained from this sample must be generalized to apply to the population. In order that these generalizations should be accurate the sample must be random, that is, every possible sample has an equal chance of selection and the choice of a member of the sample must not be influenced by previous choices, this is *simple random sampling*.

If we select members in succession from a finite population, say cards from a pack, then the chance associated with drawing a card is influenced by what cards have been drawn. The sampling is not random. If, however, each card is replaced after being drawn and the cards reshuffled, then we reduce the process to simple sampling. If the population is infinite, then we have simple sampling if we select members of the sample in succession. If the population size is large compared with the sample size, then the extraction of a relatively few members does not greatly affect the population and we regard such sampling in the same way as that from an infinite population.

The method of selection must not, however, introduce a non-random element. If an attempt is made to forecast the possible outcome of an election by choosing 10,000 people from a telephone directory and asking them how they will vote, the sample is non-random. This is because telephones are not so common among the lower income groups. A sample of apples from the top of a barrel is likely to include fewer bruised apples and thus give a much better impression of the quality of the fruit.

If a die is thrown or a coin tossed repeatedly, the results form a random sequence and are a sample of all the possible throws of the die or tosses of the coin.

Exercises 2d

The following exercises are best done in pairs with one recording the results and one performing the experiment.

1. (a) Throw one die 120 times and record the frequency with which each number occurs. What result would you have expected?

(b) Toss a coin 60 times and record the number of heads occurring. What result would you have expected?

2. Write down all possible ways in which two dice can fall if thrown together. Make a table showing the proportions in which total scores of 2, or 3, or . . . 12 can be expected. Throw two dice 72 times and record the frequency with which each possible total number occurs. What result would you have expected?

3. Continually toss a coin noting the number of consecutive heads obtained. Thus the sequence T, H, T, H, H, T, T, H, T, T, H, H, H, H,

$$\quad\quad\quad\quad\quad 1 \quad\quad 2 \quad\quad\quad 1 \quad\quad\quad 4$$

T, . . . would be noted as

Number of consecutive heads	1	2	3	4	5	6...
Frequency	2	1	–	1	–	–...

Stop when 32 sequences have been recorded. Note that the results have a different distribution to those of questions 1 or 2. Is this a random sample of such sequences?

4. Using a stop watch count the number of motor cars passing a given point on a road in each of 30 consecutive half-minute intervals. Does this give a random sample of the number of vehicles passing this point for every half minute?

5. Count the number of words in the first 100 sentences of any English reading book. Set up a frequency table. (Keep a copy of your results.)

6. Using a four-figure table of logarithms, find the logarithms of the 89 numbers 1·1, 1·2, 1·3, . . . (by steps of 0·1) . . . 9·8, 9·9. This is often written 1·1(0·1)9·9. Now add the four digits of the logarithm and if the result is less than ten record it, if the result is not less than ten add the digits of the result. Repeat until the result is a single integer and then record this integer.

Example

Number	Logarithm				Record
1·1	0·0414	9			9
...	...	...			...
...	...	...			...
...	...	...			...
...	...	...			...
1·4	0·1461	12	3		3
...	...	...	...	...	...
...	...	...	...	...	...
...	...	...	...	...	...
...	...	...	...	...	...
9·3	0·9685	28	10	1	1

7. Throw two dice and make a note of the *product* of the two numbers shown. Repeat the experiment 120 times. Set up a frequency table and keep a copy of your results.

8. Draw histograms for the data which you obtained in answer to questions 5, 6 and 7 and draw the corresponding cumulative frequency curves.

9. Draw histograms to illustrate the data of *Table 2.4*, and also the following data:

Degree of cloud (in eighths)	0	1	2	3	4	5	6	7	8
Frequency	125	47	29	20	20	22	30	60	247

2.5. RANDOM NUMBERS

From the previous sections we have seen that a sample must be random. The problem is how do we select our random sample?

Note: the word random refers to the manner in which the sample is selected rather than to the particular sample. Thus a person may thoroughly shuffle a pack of cards and randomly select four cards. If he drew four aces he has an unrepresentative sample of denominations but he still has a random sample by virtue of the manner in which it was drawn.

One way in which a random sample may be chosen is to construct a mathematical model of the population and take a sample from this model. The simplest mathematical model is one which attaches a positive integer to each member. The advantages of this method are:

(1) No physical model has to be constructed.

(2) The numbering can be carried out in any convenient way.

(3) Having set up a series of such numbers they can be applied to any enumerable population.

Tables of random numbers are available and some are reproduced at the back of this book. A lot of research has been carried out in order to ensure that such tables are, in fact, random; notably by the Rand Corporation of America. Digital computers can be programmed to generate random numbers.

Note: the same set of random numbers should not be used every time. A starting point in the table should be chosen in a random manner every time a new set of numbers is required.

2.6. HOW TO USE RANDOM SAMPLING NUMBERS

The tables at the back of this book are adequate for any exercises in the book, but there are not enough numbers to select large samples. Tables are generally given with the numbers grouped, and they can be read across or down.

When using the tables the following steps are to be followed:

(1) Decide what size numbers are required (i.e. how many digits). The minimum number that can be used will be fixed by the size of the population. More than the minimum number may be used if it makes for ease in reading the numbers from the table.

(2) Randomly choose a starting point in the tables. This can be done in several ways (throwing a die, opening a book, etc.). A neat way is to number the pages of random numbers as if they were a population and do the same for the rows and columns of the pages. Select a random number for each of these populations by means of a pin in any page. The pin is to choose the page, row and column number and must not be taken as the starting point. The method of going from the

random numbers to the number of the page, etc., is illustrated in Example 2.10.

(3) Read off a set of random numbers equal to the numbers in the sample. Do not lose the finishing point as some of the random numbers chosen may indicate the same member of the population and then more random numbers will be required to obtain a full sample with no duplication.

(4) Translate from the random numbers to the population to get the members of the population in the sample.

Note: in order to first understand rules 1, 3 and 4 we will ignore rule 2 in the following example.

Example 2.10. Choose a random sample size 5 from the following enumerable populations whose size is given. (*a*) 10, (*b*) 8, (*c*) 30, (*d*) 145. (*a*) (1) Since the population size is 10 we can use single digit random numbers with 0 corresponding to 10. (2) Assume the starting point. (3) The numbers read off in order are (say) 5, 0, 4, 7, 5. (4) These give as members of the sample 5, 10, 4, 7; the next number duplicates the first so we choose the next random number (say) 3.

The sample is 5, 10, 4, 7, 3.

(*b*) Using single random numbers it is simplest in this example to ignore those which are 8 or 9, and to let 0 correspond to the population member numbered 8.

If the random numbers are 4, 5, 9, 3, 0, 4, 2;
The sample is   4, 5,   3, 8,   2
9 is ignored and the second 4 also.

(*c*) We must take the numbers in pairs from 00 to 99. If we ignore all above 29 we will be wasting a large proportion of the numbers. In order not to waste all these numbers we allocate more than one number in the range 00–99 to each member of the population. Since we must allocate the same number of numbers to each member of the population we must find the maximum allowable. In this example there are 100 random numbers and the population size is 30. Therefore we divide 100 by 30 and ignore the remainder, i.e. we allocate 3 numbers to each member thus:

30		1		2	...	29		
00	30 60,	01	31 61,	02	32 62, ...	29	59	89

Numbers 90–99 are ignored.

A close look at this allocation of numbers will show that the remainder, on division of the random number by 30, always gives the number allocated to the population member.

Summarizing: (*i*) Divide the population size into 1 more than the maximum chosen (to allow for 0, 00, or 000, etc.), i.e. in the above example divide 30 into $99 + 1 = 100$. (*ii*) Ignoring the remainder, multiply the answer obtained in (*i*) by the population size ($3 \times 30 = 90$). (*iii*) This number and all above it will be ignored. (*iv*) Divide all random numbers by the population size. The remainder ($0 = 30$) is the member of the population chosen by the random number.

Random numbers	25	19	64	95		82	60
Remainders		25	19	4	ignore	22	0

The sample is 4, 19, 22,   25, 30.

(*d*) This part of the question is done in the manner described in part (*c*).
 $100 < 145 < 1000$, thus we take three digit random numbers 000 to 999

$$\frac{999 + 1}{145} = 6 + , \qquad 6 \times 145 = 870.$$

therefore we ignore 870 and above.

Random numbers		357	609	099	731	013
Remainders (on division by 145)		67	29	99	6	13

Sample is 6, 13, 29, 67, 99.

In Example 2.10 we assumed a starting point for our random numbers. In general we choose our starting point at random. We pin prick three numbers from the random number tables. The size of the numbers, i.e. singles, pairs, etc., depends on the number of pages of random numbers which are available, and the number of rows and columns on a page. The method detailed in Example 2.10 is used to find the page, row and column of the starting point using the three numbers obtained by a pin.

Random numbers are also useful for arranging a set of numbers in a random order.

Example 2.11. Arrange the numbers 1–8 in a random order; we proceed as for choosing a sample,

i.e. $8 < 10$ therefore use single-digit numbers.
ignore numbers 8 and 9.

Suppose from the tables, assuming we have picked our starting point in a random manner, our random numbers are: 2, 9, 9, 3, 5, 8, 2, 1, 7, 1, 0, 4. Since we cannot have duplications and 8 and 9 are ignored, we have as the random order 2, 3, 5, 1, 7, 8, 4, 6. (*Note*: 0 indicates 8.) The last number 6 automatically occurs.

Example 2.12. Arrange the numbers 1–14 in random order.

$10 < 14 < 100$ therefore we use two-digit numbers.

$$\frac{100}{14} = 7 +, \quad 7 \times 14 = 98, \text{ therefore ignore 98 and 99.}$$

Table 2.22 gives a set of random numbers chosen with a random starting point from a set of tables.

Table 2.22

Random number	10	04	00	95	85	04	32	80
Order	10	4	14	11	1	duplicate	duplicate	duplicate

Random number	19	03	29	29	80	04	21	52
Order	5	3	duplicate	duplicate	duplicate	duplicate	7	duplicate

Random number	76	23	94	97	28	60	26	48	16
Order	6	9	duplicate	13	duplicate	duplicate	12	duplicate	2

The random order is 10, 4, 14, 11, 1, 5, 3, 7, 6, 9, 13, 12, 2, 8.
Note: there is of course no need to write down all the random numbers if they are obviously duplicates.

Exercises 2e
1. Select a sample size 14 from populations with the following sizes: (a) 38, (b) 50, (c) 162, (d) 1428.
2. Arrange the numbers 1–5 in random order.
3. Arrange the numbers 1–18 in random order.
4. Choose a random sample of size 16 from the data of Exercises 2(a) question 1 and of size nine from the data of Exercises 2(a) question 2.
5. Choose a random sample of 5 members of your class. Is this a random sample of all the members of your school? Give reasons for your answer.

EXERCISES 2
1. The weights, to the nearest 0·01 of an ounce, of 80 packets of sugar delivered by an automatic packing machine were as follows:

16·05	16·21	16·13	16·25	16·20	16·19	16·20	16·26	16·23	16·28
16·21	16·19	16·24	16·10	16·11	16·15	16·13	16·15	16·09	16·19
16·17	16·21	16·10	16·20	16·20	16·12	16·26	16·22	16·27	16·28
16·21	16·15	16·17	16·16	16·12	16·24	16·07	16·26	16·18	16·12
16·21	16·14	16·23	16·22	16·07	16·24	16·22	16·22	16·18	16·31
16·08	16·23	16·12	16·22	16·16	16·16	16·18	16·22	16·30	16·19
16·17	16·23	16·28	16·25	16·17	16·29	16·16	16·11	16·16	16·27
16·29	16·13	16·17	16·22	16·14	16·29	16·19	16·30	16·19	16·32

Using equal class widths 16·045–16·095, 16·095–16·145, . . . etc., construct a frequency table of the values and draw a histogram to represent the distribution.

2. Using the results of question 1 find the total weight of the 80 packets of sugar from (*a*) the original data, (*b*) the frequency table. Compare the two answers; did you expect to obtain this result? Give reasons why it is unusual.

3. *Table 2.23* gives the relative humidity in June at various places; construct a bar chart to illustrate the data. Why is a pie chart not a suitable way of illustrating this data?

Table 2.23

Place	Relative humidity (per cent)
Brisbane	72
Calcutta	88
London	68
Aberdeen	75
Timbuktu	49
Wadi Halfa	23

4. The following table gives the rainfall, in inches, at Naples. Draw a bar chart to illustrate the data (assume each month has 30 days).

Month	Jan.	Feb.	Mar.	Apr.	May	June	July	Aug.	Sept.	Oct.	Nov.	Dec.
Rain (in)	3·4	3·0	3·0	2·8	1·8	1·3	0·8	1·2	2·9	4·3	4·9	4·2

5. Compare the populations of the following towns by a graphical method. Paris 2,900,000, Marseilles 550,000, Lyons 525,000, Bordeaux 260,000, Lille 220,000.

6. The following table gives the distribution of monthly salaries of 48 employees in an office:

Salary to nearest £	70–84	85–99	100–109	110–119	120–149	150–199	200–299
Number of employees	10	15	11	6	2	3	1

Draw a histogram to illustrate the data. Calculate, from the table, the total monthly salary bill (to the nearest £) for the 48 employees.

7. Which of the following are members of a discrete population and which are members of a continuous population?

(a) The number of motor car licences issued in Great Britain during a given month of the year?

(b) The speed of a car in miles per hour.

(c) The barometric pressures at any place, taken at noon each day.

(d) The length of a screw made by an automatic machine.

(e) The chemical analysis (percentage) of the impurities in sand which is being used for glass making.

(f) The number of petals on a flower.

(g) The total time during which an aircraft flies in a day.

8. Construct the cumulative frequency distribution of the data in question 6 and hence draw the cumulative frequency curve.

9. The male population of England and Wales in 1947 is given in Table 2.24 in thousands.

Choose an upper limit for the last class, giving reasons for your choice, and draw a histogram to show the distribution of population by age.

10. The data given in Table 2.25 relates to boys under 16 years of age on the 1st September 1948 and records their performance in an examination.

Calculate the percentage passing the examination for each age.

11. Table 2.26 gives the coal exports of Great Britain in the month of September in 1948 and 1949.

Table 2.24

Age group	Population (thousands)	Age group	Population (thousands)
0–4	1806	45–49	1425
5–9	1444	50–54	1187
10–14	1373	55–59	1076
15–19	1497	60–64	933
20–24	1567	65–69	780
25–29	1633	70–74	580
30–34	1631	75–79	340
35–39	1715	80–84	149
40–44	1629	85 and over	57

Table 2.25

Age (yrs)	Age (months)	Number taking the examination	Number passing the examination
15	11	609	454
15	10	651	500
15	9	557	472
15	8	542	418
15	7	480	369
15	6	386	287
15	5	396	310
15	4	399	328
15	3	359	292
15	2	326	260
15	1	284	179
15	0	185	165

Table 2.26

	Quantity 1948 (1000 tons)	Quantity 1949 (1000 tons)
Anthracite	52	97
Unscreened	80	76
Large	204	347
Graded	258	254
Smalls	219	279
Gas	187	147
Coking	67	8

Make two pie charts to illustrate the data. Make *one* bar chart to illustrate the data. (Keep the bars together in pairs, thus the bars representing the quantities of anthracite for 1948 and 1949 would be together, followed by the two bars for the unscreened coal, etc.) Which is the better method of representation?

12. The numbers of α-particles radiated from a disc in 2000 periods of 10 sec were as follows.

No. of α-particles	0	1	2	3	4	5	6	7	8	9	10	11
Frequency of occurrence	15	68	168	290	359	352	295	210	132	70	38	3

Draw a histogram to illustrate the data. (This is a skew distribution typical of the Poisson distribution; see Chapter 4.)

13. The following data, regarding race, sex, and conjugal status, were obtained from a study of a group of 1000 employees in a cotton mill.

525 coloured employees
312 male employees
470 married employees
42 coloured males
147 coloured married
86 married males
25 married coloured males

Examine these classifications for consistency and comment on the result of this examination. (Liv.U.)

14. Construct a histogram to illustrate the following data given in *Table 2.27*.

Table 2.27

Amount of stake on a football pool	Number of individuals
2s. 0d.– 2s. 11d.	154
3s. 0d.– 3s. 11d.	267
4s. 0d.– 7s. 11d.	198
8s. 0d.–10s. 11d.	55
11s. 0d.–15s. 11d.	27
16s. 0d. and over	11

Assuming that the stakes in any class are uniformly spread over the class, indicate by shading that part of your histogram which corresponds to individuals staking at least 5s. 0d. but less than 10s. 0d. Estimate the percentage of individuals in this class.

15. Eighty fossilized shells, selected from a coal seam, have lengths as given in *Table 2.28*:

Table 2.28

Length (cm) (to nearest 0·01 cm)	Number of shells
4·41–4·50	5
4·51–4·60	10
4·61–4·70	17
4·71–4·80	20
4·81–4·90	20
4·91–4·95	8

Draw a histogram to illustrate the data, and comment on its shape.

16. Describe the various charts and diagrams which are used for conveying statistical information.

17. *Table 2.29* gives the age distribution of the male population under 75 of the United Kingdom in 1953.

Table 2.29

Age group	Population (thousands)	Age group	Population (thousands)
Under 1	399	35–39	1710
1 and under 2	399	40–44	1890
2–4	1239	45–49	1833
5–9	2133	50–54	1593
10–14	1726	55–59	1286
15–19	1662	60–64	1081
20–24	1701	65–69	888
25–29	1780	70–74	669
30–34	1900		

Represent the above figures in the form of a histogram and make general comments on its special features. (J.M.B.)

18. Draw a histogram to illustrate the data given in *Table 2.30* and point out its main features.

Table 2.30

Size of cinema (Number of seats)	Number of cinemas
Less than 250	165
251–500	897
501–750	1043
751–1000	895
1001–1500	923
1501–2000	470
Over 2000	204
Total	4597

J.M.B. (part)

3

ELEMENTARY PROBABILITY

3.1. INTRODUCTION

THE phrases 'it is probable that', 'it is very likely that', 'there is virtually no chance' are often used in conversation. They are used to describe events which could possibly happen. They represent our estimates of the probability of the events happening. They are sometimes based on personal feelings, but more often they are based on relative frequency, i.e. the proportion of times such an event has happened previously. These subjective or intuitive ideas of probability are basic and fundamental in life and they give rise to two definitions of probability which are now discussed.

If we carry out an experiment n times and on r occasions obtain a given result, then we define the probability of obtaining the given result, when the experiment is performed once only, as the limiting value of the ratio $r:n$ as n becomes very large (tends to infinity).

This definition can only be used approximately since it is only possible to carry out a finite number of trials. Insurance companies use past records to find the probabilities of events on which they issue insurance policies. Thus the probability of a man aged 66 now, dying within the next year is taken as the ratio of the total number of men known to have died at 66 compared with the grand total of all men who died at age 66 or more. The value of the probability thus obtained is subject to modification as time goes on.

Again consider a coin; the absolute value of the probability of it coming down heads can only be obtained approximately. *Table 3.1* gives the result of tossing such a coin. The probabilities seem to be approaching the value of 0·516, but we cannot be absolutely certain.

In many probability problems a different probability definition is much more useful. Before stating this definition we must define what is meant by *equally likely* events.

Definition—If in n trials one event occurs r_1 times and a second event occurs r_2 times, we say that the events are equally likely if

$$\lim_{n \to \infty} \frac{r_1}{n} = \lim_{n \to \infty} \frac{r_2}{n} \qquad \qquad \dots (3.1)$$

32

Table 3.1

Number of throws (n)	Number of heads (r)	Probability of a head (r:n)
100	49	0·4900
200	106	0·5300
500	248	0·4960
1000	516	0·5160
10,000	5158	0·5158

We can now state our second probability definition.

Definition—If out of a total of *n* equally likely events *r* are deemed favourable, then the *probability of a favourable event* (*p*) is given by

$$p = \frac{r}{n} \qquad \qquad \dots (3.2)$$

Example 3.1. What is the probability of obtaining a score of more than four when throwing a true die?

A true die has six equally likely scores 1, 2, 3, 4, 5, or 6. Of these, two (5 and 6) are favourable

$$\text{hence} \quad \text{Pr(score} > 4) = \frac{2}{6} = \frac{1}{3}$$

Example 3.2. One egg is chosen at random from a basket containing 57 good eggs and six bad ones. What is the probability that the chosen egg is bad?

A random choice implies that each egg has an equal chance of being chosen.

Number of ways of picking an egg is 63

Number of ways of picking a bad egg is 6

Probability of picking a bad egg is $\dfrac{6}{63}$

$$\text{i.e.} \quad \text{Pr(of picking a bad egg)} = \frac{2}{21}$$

From the definition (3.2) the probability of an unfavourable event

(q) is given by

$$q = \frac{n - r}{n} \qquad \dots (3.3)$$

and since $p = \dfrac{r}{n}$ it follows that $p + q = 1$ $\dots (3.4)$

Since $0 \leqslant r \leqslant n$ the range of p is 0 to 1.

$\quad\quad$ $p = 0$ indicates an impossible event.

$\quad\quad$ $p = 1$ indicates an absolutely certain event.

Example 3.3. (*a*) The probability of a stone giving blood is 0. (*b*) The probability of a solid piece of lead sinking in water is 1.

3.2. MUTUALLY EXCLUSIVE EVENTS

Consider all the possible ways in which a card can be drawn from a pack. The 52 possibilities could be represented by a *set* (*space*) of 52 points (see *Figure 3.1*).

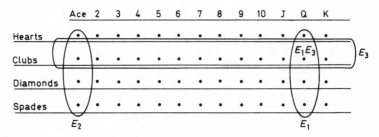

Figure 3.1

Consider the three events (indicated by rings in *Figure 3.1*) E_1, E_2, E_3 where E_1 is the probability of drawing a Queen.

$$\Pr(E_1) = \frac{4}{52} = \frac{1}{13}$$

E_2 is the probability of drawing an Ace

$$\Pr(E_2) = \frac{4}{52} = \frac{1}{13}$$

E_3 is the probability of drawing a Club

$$\Pr(E_3) = \frac{13}{52} = \frac{1}{4}$$

E_1 and E_2 have no points in common and can never occur simultaneously; that is $\Pr(E_1 \text{ and } E_2) = 0$ we say that E_1 and E_2 are *mutually exclusive events*. E_1 and E_3 have part of the set of points in common (denoted by $E_1 E_3$ and in this case consisting of only one point) that is, both E_1 and E_3 can occur simultaneously and they are not mutually exclusive.

Theorem 1. If E_1 and E_2 are two events

$$\Pr(\text{either } E_1 \text{ or } E_2) = \Pr(E_1) + \Pr(E_2) - \Pr(\text{both } E_1 \text{ and } E_2)$$
$$\dots (3.5)$$

Note that 'either E_1 or E_2' implies either one or both, that is, 'inclusive or'.

Example 3.4. Using the data of *Figure 3.1*, Pr(drawing either a Queen or a Club or both)

$$= \Pr(\text{Queen}) + \Pr(\text{Club}) - \Pr(\text{both Queen and a Club})$$

$$= \frac{1}{13} + \frac{1}{4} - \frac{1}{52}$$

$$= \frac{16}{52}$$

$$= \frac{4}{13}$$

Theorem 2. If E_1 and E_2 are mutually exclusive events

$$\Pr(\text{either } E_1 \text{ or } E_2) = \Pr(E_1) + \Pr(E_2) \qquad \dots (3.6)$$

[since $\Pr(\text{both } E_1 \text{ and } E_2) = 0$ for mutually exclusive events].

Example 3.5. Again using the data of *Figure 3.1*, Pr(drawing either a Queen or an Ace)

$$= \Pr(\text{Queen}) + \Pr(\text{Ace})$$

$$= \frac{1}{13} + \frac{1}{13}$$

$$= \frac{2}{13}$$

The theorems can be generalized for several events. For n events

$E_1, E_2, \ldots E_n$ Theorem 1 becomes:

$$\Pr(E_1 \text{ or } E_2 \text{ or } \ldots \text{ or } E_n) = \sum_{i=1}^{n} \Pr(E_i) - \sum_{i \neq j} \Pr(E_i \text{ and } E_j)$$

$$+ \sum_{i \neq j \neq k} \Pr(E_i \text{ and } E_j \text{ and } E_k) \ldots (-1)^{n-1} \Pr(E_1 \text{ and } E_2 \text{ and } \ldots$$

$$E_n) \ldots (3.7)$$

Theorem 2 becomes for n mutually exclusive events

$$\Pr(E_1 \text{ or } E_2 \text{ or } \ldots \text{ or } E_n) = \Pr(E_1) + \Pr(E_2) \ldots + \Pr(E_n).$$

$$\ldots (3.8)$$

3.3. INDEPENDENT EVENTS

The previous section was concerned with a *single* trial and the probabilities that could be associated with it. In this section we consider *two* or *more* trials and the probabilities associated with the results of a combination of trials.

We say that two or more events are *statistically independent* if the probability of either one of the events is unaltered by the occurrence or non-occurrence of the other event.

In the case of drawing two cards from a pack of cards, if the first card is not replaced in the pack before the second card is drawn then the drawing of the first card alters the situation for the drawing of the second card. For example, the probability of drawing an Ace on the second draw is not $\frac{1}{13}$ if the first card is not replaced. It is either $\frac{3}{51}$ or $\frac{4}{51}$, depending on whether or not an ace was obtained on the first draw.

Theorem 3. If $E_1, E_2, \ldots E_n$ are n statistically independent events then

$$\Pr(E_1, E_2, E_3, \ldots E_n \text{ all occurring}) = \Pr(E_1) \times \Pr(E_2) \ldots \times \Pr(E_n)$$

$$\ldots (3.9)$$

Example 3.6. A card is drawn from a well-shuffled pack and a true die is thrown. What is the probability of drawing an Ace and throwing a six?

$$\Pr(\text{Ace}) = \frac{1}{13}, \quad \Pr(\text{Six}) = \frac{1}{6}.$$

The two events are statistically independent

$\therefore$ Pr(both an Ace and a Six) = Pr(Ace) $\times$ Pr(Six)

$$= \frac{1}{13} \times \frac{1}{6}$$

$$= \frac{1}{78}$$

Example 3.7. A case contains twelve valves, four of which are known to be defective. One valve is chosen at random, tested and replaced; another is then chosen at random and tested. What is the probability of drawing two defective valves? What are the other possible results of the two choices?

Let B be a defective valve, and G be a non-defective valve.

On the first choice $\quad$ Pr(B) $= \dfrac{4}{12} = \dfrac{1}{3}$.

Since the first choice is replaced, for the second choice we also have

$$Pr(B) = \frac{1}{3}$$

$$\therefore Pr(B \text{ and } B) = \frac{1}{3} \times \frac{1}{3} = \frac{1}{9} \qquad \ldots (i)$$

Note: in this example the act of replacement has made the two events statistically independent.

The other possible mutually exclusive results are BG, GB, GG.

$$\text{Now} \quad Pr(G) = \frac{8}{12} = \frac{2}{3}$$

$$\therefore Pr(B \text{ and } G) = \frac{1}{3} \times \frac{2}{3} = \frac{2}{9} \qquad \ldots (ii)$$

$$Pr(G \text{ and } B) = \frac{2}{3} \times \frac{1}{3} = \frac{2}{9} \qquad \ldots (iii)$$

$$Pr(G \text{ and } G) = \frac{2}{3} \times \frac{2}{3} = \frac{4}{9} \qquad \ldots (iv)$$

Having considered all possible outcomes of the two choices the sum of their probabilities must be unity. Adding the results (i) ... (iv)

we have $\dfrac{1}{9} + \dfrac{2}{9} + \dfrac{2}{9} + \dfrac{4}{9} = 1.$

Theorem 4. The sum of the probabilities associated with all possible mutually exclusive events which are relevant to a given situation is unity.

If the events are $E_1, E_2, \ldots E_n$ and $\Pr(E_i) = p_i$ (all i) then $\sum\limits_{i=1}^{n} p_i = 1$

Example 3.8. Three boxes each contain 1000 resistors. The number of defective resistors in the three boxes are respectively 100, 50 and 80. One resistor is drawn from each box; assuming every resistor has an equal chance of being chosen, what is the probability that of the three resistors at least one is defective?

The three events (drawing a resistor) are statistically independent and the three probabilities of drawing a defective resistor are:

$$\text{From the first case} \quad \frac{100}{1000} = \frac{1}{10}$$

$$\text{From the second case} \ \frac{50}{1000} = \frac{1}{20}$$

$$\text{From the third case} \quad \frac{80}{1000} = \frac{2}{25}$$

Since the question asks for the probability of at least one we require the sum of the probabilities of drawing one, two or three. However, since in this case there is only one other mutually exclusive event, namely, no defectives, it is simpler to use

$$1 - \Pr(\text{drawing no defectives})$$

to obtain the probability of at least one defective.
Probabilities of not drawing a defective are:

$$\text{First case} \quad \frac{900}{1000} = \frac{9}{10}$$

$$\text{Second case} \ \frac{950}{1000} = \frac{19}{20}$$

$$\text{Third case} \quad \frac{920}{1000} = \frac{23}{25}$$

$$\text{Therefore} \quad \Pr(\text{drawing no defectives}) \quad = \frac{9}{10} \times \frac{19}{20} \times \frac{23}{25}$$

and $\quad$ Pr(drawing at least one defective) $= 1 - \dfrac{9}{10} \times \dfrac{19}{20} \times \dfrac{23}{25}$

$$= 1 - \dfrac{3933}{5000}$$

$$= \dfrac{1067}{5000}$$

$$= 0{\cdot}2134.$$

Note: in certain problems events are assumed to be independent when in actual fact they are not. For example, the probability of drawing a defective nut from a case containing 10,000 of which 50 are defective is $\dfrac{50}{10,000}$, i.e. $\dfrac{1}{200}$. If the nut is not replaced the probability of drawing a second defective nut is either $\dfrac{50}{9999}$ or $\dfrac{49}{9999}$. The difference between these three values can generally be regarded as insignificant and all three probabilities are taken as $\dfrac{1}{200}$. Thus the probability of drawing two defectives is $\dfrac{1}{200} \times \dfrac{1}{200} = \dfrac{1}{40,000}$.

Example 3.9. In a case containing 10,000 bolts 5 per cent are either oversize or undersize. If three bolts are picked at random what is the probability that all three are within the required limits?

If 5 per cent are either undersize or oversize
95 per cent are within the required limits
$$\text{Pr(correct bolt)} = \dfrac{19}{20}$$

Treating the three choices as independent

$$\text{Pr(picking three correct bolts)} = \dfrac{19}{20} \times \dfrac{19}{20} \times \dfrac{19}{20}$$

$$= \dfrac{6859}{8000}$$

$$= 0{\cdot}857$$

To avoid confusion we would stress that:

(1) *Mutual exclusion* is considered when there are several possible outcomes of *one* action.

(2) *Independence* is considered when there are *two* or *more* actions which can occur together.

c

Exercises 3a

1. If two true dice are thrown together, which of the following pairs of events are mutually exclusive:

(a) The two numbers thrown are the same: at least one 4 is thrown.

(b) The total score is 5: the total score is 6.

(c) At least one 6 is thrown; the total score is less than 6.

(d) At least one 5 is thrown: the total score is more than 6.

2. Which of the following pairs of events are independent?

(a) A battery is chosen for testing from a case containing 100. Another battery is chosen for testing from another case containing 100.

(b) A battery is chosen from a case; another battery is chosen from the same case.

(c) A die is thrown and a pack of cards is cut.

3. Illustrate in a diagram the following events:

(a) E_1, E_2, E_3 all mutually exclusive events.

(b) E_1 and E_2 mutually exclusive
E_1 and E_3 mutually exclusive
E_2 and E_3 *not* mutually exclusive.

(c) E_1 and E_2 mutually exclusive
E_1 and E_3 *not* mutually exclusive
E_2 and E_3 *not* mutually exclusive.

(d) E_1 and E_2 *not* mutually exclusive
E_1 and E_3 *not* mutually exclusive
E_2 and E_3 *not* mutually exclusive.

4. Two true dice are thrown; find the probabilities that the total score is (a) over 10, (b) under 5, (c) 13, (d) 1, (e) an even number.

5. Consider any of the following experiments:

(a) Tossing a true coin and noting the number of heads and tails.

(b) Throwing two dice and noting the total score.

(c) Drawing a card from a well-shuffled pack and noting the suit. Perform the experiment 200 times, recording the results. On the basis of your results what is the probability for each event? What is the theoretical probability for each event?

6. Three true coins are tossed; what is the expected (theoretical) probability of (a) all heads, (b) at least one tail?

7. Toss four coins 100 times noting the number of times three heads are shown. Does the observed probability agree with the expected (theoretical) probability (assuming the coins are true)?

8. Repeat question 7 using two dice and noting the number of times two sixes are shown.

9. A pile of coins consists of 12 halfpennies, 12 pennies, 12 sixences, 12 shillings and 12 halfcrowns. One coin is selected, its value noted and then returned to the pile. Another coin is then selected and its

value noted. Assuming every coin has an equal chance of being chosen what are the probabilities that:

(a) A copper coin is chosen as the first coin?

(b) A silver coin is chosen as the second coin?

(c) The first coin is a halfcrown and the second coin is a silver coin?

(d) Two copper coins are chosen in succession.

10. A and B toss a true coin, the first to throw a tail wins. If A starts what are their respective chances of winning?

3.4. INTRODUCTION TO PERMUTATIONS AND COMBINATIONS

Referring back to our definition of probability (equation 3.2) we see that it is necessary to be able to enumerate all 'favourable' and all possible events in order to find the probability of a favourable event. Permutations and combinations deal with the grouping and arrangement of objects and are very useful in calculating probabilities. It is necessary to start with the following theorem.

Theorem 5. If an event can take place in w_1 ways and when this has happened a second event can take place in w_2 ways, the number of ways in which the two events can take place is $w_1 \times w_2$ ways.

Example 3.10. A jam manufacturer has the choice of three different kinds of sugar and four different grades of strawberries. In how many ways can he choose the sugar and the strawberries?

From the theorem   Number of ways = $3 \times 4 = 12$.

The theorem can be extended to more than two events and for n events the total number of ways is $w_1 \times w_2 \times w_3 \times \ldots \times w_n$ ways.

Example 3.11. A number lock has five rings each with nine different digits $1, 2, \ldots 9$. How many possible five-figure numbers can be made?

The first ring can be set in nine ways, and when this has been done the second ring can be set in nine ways. Similarly for the third, fourth and fifth rings.

$$\text{Total number of ways} = 9 \times 9 \times 9 \times 9 \times 9$$
$$= 59,049 \text{ ways}$$

Exercises 3b

1. How many groups of one consonant, one vowel and one number, in this order, can be made from '123 East Rd'. Write out all the groups, e.g. R a 1, t E 2, etc.

2. The result of a football match can be either win, lose or draw; how many possible series of results are there for four matches?

3. In an election for the offices of secretary and treasurer there are four candidates, A, B, C, and D. In how many ways is it possible to vote? Write out all the possible ways of voting (secretary first, treasurer second), e.g. AB, BA, AC, . . . , etc.

4. If in question 3 the election had been for two members of a committee from the four candidates (A, B, C, D) write out all possible ways of casting the two votes. Why is the answer different from that of question 3?

5. Three people enter a railway carriage in which there are six seats. In how many ways can they each occupy a seat?

Permutations

In questions 3 and 4 in the preceding Exercises 3(b) we had two different situations. In question 3 the *order* of things was important but in question 4 it was not. This is the difference between permutations and combinations.

Definition—Each ordering or arrangement of all or part of a set of objects is called a *permutation*. For example, all possible permutations of all of the three letters ABC are ABC, ACB, BAC, BCA, CAB, CBA. We can also consider permutations of sets of objects which form part of a parent group. For example, the possible permutations of any three of the four letters A, B, C, D (written $^4P_3$) are:

$$
\begin{array}{cccccc}
\text{ABC} & \text{ACB} & \text{BAC} & \text{BCA} & \text{CAB} & \text{CBA} \\
\text{ABD} & \text{ADB} & \text{BAD} & \text{BDA} & \text{DAB} & \text{DBA} \\
\text{ACD} & \text{ADC} & \text{CAD} & \text{CDA} & \text{DAC} & \text{DCA} \\
\text{BCD} & \text{BDC} & \text{CBD} & \text{CDB} & \text{DBC} & \text{DCB}
\end{array}
$$

. . . . (3.10)

that is $^4P_3 = 24$.

A general formula can be obtained from n things taken r at a time (written ^nP_r) $(r \leqslant n)$ by making use of our *Theorem 5*.

From n things the first can be chosen in n ways and when this has been done there are $n - 1$ things left. So the second can be chosen in $n - 1$ ways. From our theorem the first *two* can be chosen in $n(n - 1)$ ways.

There are now $n - 2$ things left, so the third thing can be chosen in $n - 2$ ways; thus the first *three* can be chosen in $n(n - 1)(n - 2)$ ways.

It is clear that continuing this process we have

$$\overset{\longleftarrow\text{--------}r\text{ brackets}\text{--------}\longrightarrow}{^nP_r = n(n - 1)(n - 2)(n - 3)\ldots(n - r + 1)} \quad \ldots\text{. (3.11)}$$

Referring back to our array (3.10) the number of permutations of four things three at a time is given by

$$\overset{\longleftarrow\text{3 brackets}\longrightarrow}{^4P_3 = (4)(3)(2) = 24}$$

In the special case when permutations are made of n things n at a time $^nP_n = n(n - 1).\,(n - 2)\ldots3.2.1$. This product of all the natural numbers from 1 to n is written $n!$ and is called *factorial n*. Hence $^nP_n = n!$

Note that we can rewrite equation (3.11)

$$^nP_r = \frac{n(n - 1)(n - 2)\ldots(n - r + 1)(n - r)(n - r - 1)\ldots3.2.1}{(n - r)(n - r - 1)\ldots3.2.1}$$

$$= \frac{n!}{(n - r)!} \qquad (r \leqslant n) \qquad \qquad \ldots\,(3.12)$$

Note: in order that equation (3.12) will apply for all values of r between 0 and n we define $0! = 1$,

$$\text{i.e. } ^nP_n = \frac{n!}{(n - n)!} = \frac{n!}{0!} = n!$$

Example 3.12. In a coach with ten passenger seats in how many different ways can six passengers each occupy a seat?

This is a question of permutations of six things chosen from ten. Therefore the number of ways is

$$\overset{\longleftarrow\text{6 terms}\longrightarrow}{^{10}P_6 = 10.9.8.7.6.5}$$
$$= 151,200$$

It can be seen that the arithmetic can be quite lengthy and sometimes if the answer is only required to four or five figure accuracy we can use equation (3.12) and tables of log factorials.

Example 3.13. Find the value of $^{30}P_{19}$.

$$^{30}P_{19} = \frac{30!}{(30 - 19)!}$$

	No.	Log
	30!	32·4237
$= \dfrac{30!}{11!}$	11!	7·6012
		24·8225

$$= 6\cdot346 \times 10^{24}$$

Exercises 3c
1. (*a*) Evaluate $^5P_2$, $^6P_3$, $^7P_7$, $^6P_4$.

(b) Evaluate using log factorials $^{25}P_{17}$, $^{36}P_{15}$, $^{29}P_{10}$, $^{210}P_{67}$.

(c) Rewrite using factorials $(n + 1)(n!)$, $9 \times 7 \times 5 \times 3 \times 1$.

(d) Show that $\dfrac{20!}{10!} = 38.34.30.26 \ldots 10.6.2$.

2. In a newspaper competition ten different motor car accessories are to be put into order of usefulness. In how many different ways can they be placed? If the entry fee was 3d. per attempt and assuming sufficient entries were received to include every possible permutation once only, what was the total amount received?

3. In how many ways can a secretary, treasurer, president and vice-president be chosen from eight candidates?

4. If in question 3 three of the candidates were only eligible for the offices of president and vice-president and the other five were only eligible for the offices of secretary and treasurer, in how many ways could the four posts be filled?

5. There are eight bells in a church steeple. If each bell is rung once and once only, in how many different orders can the eight bells be rung?

Combinations

In the case of *combinations* the order is not important, so that ABC, ACB, BAC, BCA, CAB, and CBA are six different permutations but are all the same combination. The number of combinations of four things taken three at a time (written $^4C_3$) is four, namely

$$\begin{array}{ll} & \text{ABC} \\ & \text{ABD} \\ & \text{ACD} \\ & \text{BCD} \end{array} \qquad \ldots (3.13)$$

Thus $^4C_3 = 4$.

Comparing the combinations given in (3.13) with the permutations given in (3.10) we see that each combination of three gives rise to six different permutations. These arise because in permutations we can write the three letters in every possible order,

i.e. $^3P_3$ or 3! ways.

Thus the number of combinations in (3.13) multiplied by 3! is equal to the number of permutations in (3.10) or

$$^4C_3 \times 3! = {}^4P_3$$

Generalizing this idea, the number of combinations of n things taken r at a time is ^nC_r. Each of these combinations of r things can be rearranged in $r!$ ways to give all the permutations of n things taken r

at a time, hence

$$^nC_r \times r! = {}^nP_r$$

or
$$^nC_r = \frac{^nP_r}{r!}$$

Now from equation (3.11) $\overset{\longleftarrow \text{---} r \text{ brackets} \text{---} \longrightarrow}{{}^nP_r = n(n-1)(n-2)\ldots(n-r+1)}$

$$\therefore {}^nC_r = \frac{\overset{\longleftarrow \text{---} r \text{ brackets} \text{---} \longrightarrow}{n(n-1)(n-2)\ldots(n-r+1)}}{r!}$$

$$\ldots (3.14)$$

From equation (3.12)
$$^nP_r = \frac{n!}{(n-r)!}$$

$$\therefore {}^nC_r = \frac{n!}{r!(n-r)!} \qquad \ldots (3.15)$$

Equation (3.14) is useful for arithmetical calculation and equation (3.15) for algebraic manipulation.

Example 3.14. In how many ways can a hand of three cards be drawn from a pack of 52 different cards?

This is a combination, since the order in which the cards are dealt does not matter (Ace, King, Queen is the same hand whatever the order in which the cards were received).

The number of ways is $^{52}C_3$ and from equation (3.14)

$$^{52}C_3 = \frac{52.51.50}{3.\ 2.\ 1}$$

$$= 22,100$$

Example 3.15. Find the value of $^{15}C_{11}$

$$^{15}C_{11} = \frac{15.14.13.12.11.10.9.8.7.6.5}{11.10.\ 9.\ 8.\ 7.\ 6.5.4.3.2.1}$$

$$= 1365$$

We notice that in the preceding example there was a large amount of cancellation. In fact

$$^{15}C_{11} = \frac{15.14.13.12}{4.\ 3.\ 2.\ 1}$$

$$= {}^{15}C_4$$

$$= {}^{15}C_{(15-11)}$$

In general, considering equation (3.15), we have that

$$^nC_r = \frac{n!}{r!(n-r)!}$$

$$= \frac{n!}{(n-r)!r!} \quad \text{but } r = n - (n-r)$$

$$= \frac{n!}{(n-r)![n-(n-r)]!}$$

$$^nC_r = {}^nC_{n-r} \qquad \ldots (3.16)$$

Thus we can avoid cancelling when $r > \dfrac{n}{2}$ by using equation (3.16).

Example 3.16. Evaluate $^{15}C_{10}$.

$$^{15}C_{10} = {}^{15}C_{15-10}$$

$$= \frac{15.14.13.12.11}{5.\ 4.\ 3.\ 2.\ 1}$$

$$= 3003$$

Example 3.17. A customer buys 20 resistors and tests a sample of four of them;
(*a*) In how many ways can he pick the sample of four?
(*b*) If three out of the 20 resistors are faulty, how many of the possible samples of size four will include at least one faulty resistor?
(*a*) The order in which the four resistors are picked is not important therefore the number of ways is

$$^{20}C_4 = \frac{20.19.18.17}{4.\ 3.\ 2.\ 1} = 4845 \qquad \ldots (i)$$

(*b*) We now have three possible cases, namely the number of samples containing 1, 2 or 3 faulty resistors respectively.
(*i*) *Exactly one faulty resistor.* The other three resistors must be good and these can be any combination of three from 17 good ones, i.e. $^{17}C_3$; the faulty ones can be any one of the three faulty ones, i.e. $^3C_1$. Hence the required number is

$$^{17}C_3 . {}^3C_1 = \frac{17.16.15}{3.\ 2.\ 1} . \frac{3}{1}$$

$$= 2040 \qquad \ldots (ii)$$

(ii) *Exactly two faulty resistors.* Here we pick two good resistors from 17, combined with two faulty resistors from three.

$$\text{Required number is } {}^{17}C_2 \cdot {}^{3}C_2 = \frac{17.16}{1.\ 2}\ \frac{3.2}{2.1}$$

$$= 408 \qquad \ldots\text{(iii)}$$

(iii) *Exactly three faulty* resistors. The required number is

$$ {}^{17}C_1 \cdot {}^{3}C_3 = \frac{17}{1} \cdot \frac{3.2.1}{3.2.1} $$

$$= 17 \qquad \ldots\text{(iv)}$$

The sum of equations (ii), (iii) and (iv) is the required answer, which is

$$2040 + 408 + 17 = 2465 \qquad \ldots\text{(v)}$$

Note that if we now calculate the number of samples of four containing no faulty resistors, which is

$$ {}^{17}C_4 = \frac{17.16.15.14}{4.\ 3.\ 2.\ 1} = 2380 \qquad \ldots\text{(vi)}$$

this total together with that of (v) gives all the possible occurrences and equals the value (i). In fact, the best way to work out the answer to part (b) is to calculate the number of samples containing no faulty resistors (vi) and subtract this from the total number of possible samples (i).

$$\text{Required number} = \text{(i)--(vi)}$$
$$= 4845\text{--}2380$$
$$= 2465 \text{ as before}$$

Exercises 3d

1. (a) Calculate ${}^{15}C_5$, ${}^{24}C_{21}$, ${}^{9}C_4$.
(b) If ${}^{n}C_{11} = {}^{n}C_5$ find n.
(c) Show that (i) ${}^{7}C_2 + {}^{7}C_3 = {}^{8}C_3$; (ii) ${}^{11}C_5 + {}^{11}C_6 = {}^{12}C_6$.

2. A bag contains seven white and four black balls. In how many ways can (a) four balls be drawn from the bag? (b) four white balls be drawn from the bag? (c) four balls be drawn from the bag including at least one black one?

3. Two cards are drawn from a pack. In how many ways (a) can they be drawn? (b) can two Aces be drawn?
Find the probability that both cards are Aces.

4. A committee of ten is to be selected from three accountants,

eight managers and six scientists. If the committee is to consist of one accountant, five managers and four scientists how many different committees are possible?

5. A box contains 30 electric light bulbs of which six are faulty. In how many ways can: (a) a sample of five be picked? (b) a sample of five good bulbs be picked? (c) a sample of five be picked if it is to contain less than three faulty bulbs?

3.5. PROBABILITY DISTRIBUTIONS

In example 3.17 we calculated the number of ways in which samples of four could occur with 0, 1, 2 or 3 faulty resistors. If we divide each of these numbers by the total number of possible samples we obtain the probabilities of a sample containing 0, 1, 2 or 3 faulty resistors (defectives) as shown in *Table 3.2*.

Table 3.2

No. of faulty resistors	No. of samples	Probability
0	2380	$\dfrac{2380}{4845} = 0\cdot491$
1	2040	$\dfrac{2040}{4845} = 0\cdot421$
2	408	$\dfrac{408}{4845} = 0\cdot084$
3	17	$\dfrac{17}{4845} = 0\cdot003$

Such a table gives the *probability distribution* for the various samples. Thus for any set of mutually exclusive events the probabilities associated with the events give a probability distribution; in this case a discrete probability distribution.

Example 3.18. Find the probabilities associated with all the possible scores when two true dice are thrown.

The total number of ways in which two dice can fall is $6 \times 6 = 36$ and since the dice are assumed 'true' (and hence each has 6 equally likely scores 1, 2, ... 6) the probabilities of the various scores 2, 3, 4, ... 12 are worked out in *Table 3.3*.

Table 3.3

Score (x)	2	3	4	5	6	7	8	9	10	11	12
No. of possible ways	1	2	3	4	5	6	5	4	3	2	1
Probability (p)	$\frac{1}{36}$	$\frac{2}{36}$	$\frac{3}{36}$	$\frac{4}{36}$	$\frac{5}{36}$	$\frac{6}{36}$	$\frac{5}{36}$	$\frac{4}{36}$	$\frac{3}{36}$	$\frac{2}{36}$	$\frac{1}{36}$

This is a discrete probability distribution. It can be represented by a histogram, as shown in *Figure 3.2*.

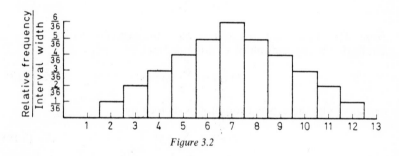

Figure 3.2

Exercises 3e

1. Find the probabilities associated with all the possible scores when three true dice are thrown together. Draw a histogram to represent this discrete probability distribution.

2. In the previous question the number of possible ways in which scores of 3, 4, 5, 6, 7, and 8 can be obtained form a progression with a definite law of formation. What is this law? If the law held for scores of 9 and 10, what would be the number of possible ways of throwing 9 and 10? Account for the discrepancy.

3. All the Kings, Queens and Jacks are removed from two packs of playing cards. Each pack is cut at random and the total score noted (an Ace counts as one). Find the probabilities associated with all the possible scores 2, 3, 4, . . . 20. Draw a histogram to represent this discrete probability distribution.

4. Five true coins are tossed and the number of heads noted. Find the probabilities associated with obtaining 5, 4, . . . 0 heads. Draw a histogram to represent this discrete probability distribution.

3.6. MATHEMATICAL EXPECTATION AND ARITHMETIC MEAN

If a man has a probability of 0·25 of winning a prize of £8 we would say that his 'chance' is worth £0·25 × 8 = £2. In general, if p is his probability of winning a prize worth £x his 'chance' would be worth £px. This leads us to define mathematical expectation as px and to extend this idea to n discrete variables as follows:

Definition. Suppose that we have n mutually exclusive events x_1, x_2, . . . x_n each with associated probabilities p_1, p_2, . . . p_n the mathematical expectation of x, written $\mathscr{E}(x)$ is defined by

$$\mathscr{E}(x) = \sum_{i=1}^{n} p_i x_i \qquad \ldots (3.17)$$

Note: $\mathscr{E}(x)$ is generally just referred to as the *expectation* of x. Similarly the expectation of x^2 is defined by

$$\mathscr{E}(x^2) = \sum_{i=1}^{n} p_i x_i^2 \qquad \ldots (3.18)$$

and generally the expectation of x^n is defined by

$$\mathscr{E}(x^n) = \sum_{i=1}^{n} p_i x_i^n \qquad \ldots (3.19)$$

The following three results apply to expectations.
(a) $\mathscr{E}(a) = a$ (a constant)

$$\text{By definition} \quad \mathscr{E}(a) = \sum_{i=1}^{n} p_i a$$

$$= a \sum_{i=1}^{n} p_i$$

$$= a \quad (\sum_{i=1}^{n} p_i = 1)$$

(b) $\mathscr{E}(ax) = a\mathscr{E}(x)$ (a constant)

$$\text{By definition} \quad \mathscr{E}(ax) = \sum_{i=1}^{n} p_i ax$$

$$= a \sum_{i=1}^{n} p_i x = a\mathscr{E}(x)$$

$(c)\ \mathcal{E}(x + y) = \mathcal{E}(x) + \mathcal{E}(y)$

By definition $\mathcal{E}(x + y) = \sum_{i=1}^{n} p_i(x_i + y_i)$

$$= \sum_{i=1}^{n} p_i x_i + \sum_{i=1}^{n} p_i y_i$$

$$= \mathcal{E}(x) + \mathcal{E}(y).$$

The *arithmetic mean* of a set of N values is obtained by adding them all together and dividing by N. It is generally just referred to as the *mean*.

Thus
$$\text{Mean} = \frac{\sum_{i=1}^{N} x_i}{N}$$

If the x_i's are repeated with frequencies f_i we have $N = \sum_{i=1}^{n} f_i$ and

$$\text{Mean} = \frac{\sum_{i=1}^{n} f_i x_i}{N} \qquad \dots (3.20)$$

Note: In future when the range of summation is obvious the suffix 'i' is omitted and, for example, the Mean written

$$\frac{\sum f x}{N}$$

Population Mean

As we remarked in section 2.4, in general only a sample of the population can be considered and so we have to distinguish between μ, the mean of the whole population, and $\bar{x}$ the mean of the sample.

Consider first
$$\mu = \frac{\sum_{i=1}^{n} f_i x_i}{N} \qquad \begin{array}{l} (x_i \text{ being taken throughout} \\ \text{the whole population)} \end{array}$$

This may be written as $\mu = \sum_{i=1}^{n} \frac{f_i}{N} x_i,$

and recalling that $\dfrac{f_i}{N}$ is the definition of the probability p_i associated with x_i [see equation (3.2)] we see that:

$$\mu = \sum_{i=1}^{n} p_i x_i \qquad \begin{array}{l} (p_i \text{ and } x_i \text{ being taken throughout the whole} \\ \text{population)} \end{array}$$

i.e. $\mu = \mathcal{E}(x)$ $\qquad \dots (3.21)$

Note: this holds only when we are dealing with a complete population.

Sample Mean

If $x_1, x_2, \ldots x_n$ is a sample of size n from a population which has mean μ and size n, then each of $x_1, x_2, \ldots x_n$ is a member of the population and

$$\mathscr{E}(x_j) = \mu \quad (x_j \text{ being any member of the sample}).$$

Let $\bar{x}$ be the sample mean, then

$$\bar{x} = \sum_{j=1}^{n} x_j/n$$

$$\therefore \mathscr{E}(\bar{x}) = \mathscr{E}(\sum_{j=1}^{n} x_j/n)$$

$$= \frac{1}{n} \mathscr{E}\left(\sum_{j=1}^{n} x_j \right) \quad [\text{see section 3.6}(b)]$$

$$= \frac{1}{n} \sum_{j=1}^{n} \mathscr{E}(x_j) \quad [\text{see section 3.6}(c)]$$

$$= \frac{1}{n} \sum_{j=1}^{n} \mu$$

$$= \frac{1}{n} . n\mu$$

$$\therefore \mathscr{E}(\bar{x}) = \mu \qquad \qquad \ldots . (3.22)$$

Since both $\mathscr{E}(x_j)$ and $\mathscr{E}(\bar{x})$ are μ, the sample mean and all sample members are said to be unbiased estimators of μ.

Example 3.19. Find μ for the data of *Table 3.3*.
From the table and equation (3.17) we have that:

$$\mathscr{E}(x) = \frac{1}{36} \times 2 + \frac{2}{36} \times 3 + \frac{3}{36} \times 4 \ldots + \frac{1}{36} \times 12$$

$$= \frac{2 + 6 + 12 + 20 + 30 + 42 + 40 + 36 + 30 + 22 + 12}{36}$$

$$= \frac{252}{36}$$

$$= 7$$

$$\therefore \mu = \mathscr{E}(x) = 7.$$

Example 3.20. Two dice were thrown together. The total score was noted each time and the results for 108 throws were:

Total score	2	3	4	5	6	7	8	9	10	11	12
Frequency	2	6	8	11	16	17	15	13	11	5	4

Find the mean of these results and discuss whether it is a sample mean or a population mean.

From equation (3.20) we have that

$$\text{Mean} = \frac{\sum fx}{N}$$

$$= \frac{2 \times 2 + 6 \times 3 + 8 \times 4 + \dots + 5 \times 11 + 4 \times 12}{108}$$

$$= \frac{774}{108}$$

$$= 7{\cdot}16\dot{}$$

In this case the 108 throws are a *sample* of all such possible throws of the two dice and thus the result is a *sample mean* and we write $\bar{x} = 7{\cdot}17$ (correct to two decimal places).

It is not possible to find the actual population mean (μ) by such experiments, we can only estimate it, and from the above results the best estimate we have of the value of μ is $7{\cdot}17$.

In the previous Example 3.19 we found μ, the population mean, because we made the assumption (see Example 3.18 from whence the data was taken) that *true* dice had been used and we found the $\mathscr{E}(x)$ which is μ, the population mean.

Example 3.21. A prize of £100 is offered for naming the three winning horses, in correct order, in a race for which ten horses are entered. What is the expectation of an uninformed person?

In this case we have only one value of the variate, namely $x_1 = £100$. The number of ways of picking three horses, in order, from ten is $^{10}P_3$ which is 720. Therefore the probability of x_1 is $p_1 = \dfrac{1}{720}$.

$$\mathscr{E}(x) = p_1 x_1 \text{ (only one value of } x)$$

$$= \frac{1}{720} \times £100$$

$$= 33\tfrac{1}{3}\text{d.}$$

$$= 2\text{s. } 9\tfrac{1}{3}\text{d.}$$

Exercises 3f

1. If a single true die is thrown, write down the probability distribution of the scores $x = 1, 2, 3, 4, 5, 6$ and obtain $\mathscr{E}(x)$ for the distribution.

2. If three coins are tossed together, write down all the possible arrangements of heads and tails. Assuming the coins are true, obtain the theoretical probabilities of 0, 1, 2 or 3 heads. Find the mean of the distribution.

3. If in question 2 four true coins had been tossed instead of three, find the mean of the probability distribution.

4. Toss three coins together and note the number of heads. Repeat the experiment 60 times and find the mean of your results. Is the answer necessarily the same as that of question 2? Give reasons for your answer.

5. Repeat question 4 using four coins. Is the answer necessarily the same as that of question 3? Give reasons for your answer.

EXERCISES 3

1. Three true dice are thrown together.

(*a*) In how many ways can they fall?

(*b*) In how many ways can they fall so that the total score is five?

(*c*) What is the probability that the total score is five?

2. A coin and a true die are thrown together. What is the chance of a head and a five turning up? Are we dealing with independent events?

3. Five cards are dealt from a pack of playing cards. If the order in which they are dealt does not matter, in how many different ways can they be dealt?

4. (*a*) If $3 \cdot {}^{2n}C_3 = 44 \cdot {}^nC_2$ find the value of n.

(*b*) Two numbers a and b can with equal probabilities have any of the values 0, 1, 2, ... 9. Find the probability that (*i*) The sum is greater than 12, (*ii*) The difference between them, irrespective of sign, is greater than five.

5. The chances of four runners in a race breaking the record are $\frac{1}{3}, \frac{3}{4}, \frac{1}{2},$ and $\frac{3}{5}$ respectively. What is the probability that the record will be broken?

6. How many permutations can be made using all the letters in the word 'education'? How many if only the vowels are used?

7. A golfer using the correct club has a probability of $\frac{1}{3}$ of making a good shot. If he uses an incorrect club he has a probability of $\frac{1}{4}$ of

making a good shot. He carries six clubs, only one of which is correct for a particular shot. He chooses a club at random.

(a) What is the probability that he makes a good shot?

(b) Check the answer to part (a) by finding the probability that he makes a bad shot $(\Pr(a) + \Pr(b) = 1)$ (there are two possible, mutually exclusive ways of making a good shot, viz. (i) right club, good shot (ii) wrong club, good shot).

8. The probabilities that two independent events A and B will occur are p_1 and p_2 respectively. What are the probabilities that:

(a) Both A and B occur?

(b) A occurs but not B?

(c) B occurs but not A?

(d) Neither A nor B occurs?

9. Each of two bags contains eight coins. How many different combinations of six coins can be made by drawing three coins out of each bag when: (a) no two of the 16 coins are alike? (b) two of the coins in *one* bag are alike.

In case (a) what is the probability that a specified coin will appear in any one combination?

10. (a) Three balls are drawn at random from a bag containing three red, four white and five black balls. Calculate the probabilities that the three balls are (i) all black; (ii) one white, one red, one black.

(b) The independent probabilities that three components of a television set will need replacing within a year are $\frac{1}{10}, \frac{1}{12}$ and $\frac{1}{15}$. Calculate the probability that: (i) at least one component will need replacing. (ii) one and only one component will need replacing.

11. In a large batch of components 2·5 per cent are defective. The batch is so large that the probability of obtaining a defective component can be considered to be unaltered by sampling. What is the chance of getting at least one defective in samples of 5, 10, 50 and n respectively?

12. In a game of cards, five cards are dealt to each player. The order in which the cards are dealt is of no importance. Find the number of possible hands containing: (a) Five cards of the same suit. (b) Four cards of one suit and the other card of a different suit. (c) Five cards of the same suit in sequence. (d) Five cards in sequence whether of the same suit or not.

13. A very long convoy of military vehicles, each of which is 18 ft long, comes to rest with a distance of 60 ft between the fronts of any two consecutive vehicles. What is the probability that the 12 ft wide drive of a house on the side of the road is wholly or partially blocked?

14. (a) A man purchases a sweepstake ticket. He can win a first prize of £5000, a second prize of £1000 or a third prize of £100 with

probabilities 0·0001, 0·0004, and 0·001 respectively. What is a fair price to pay for the ticket? (Calculate his expectation.)

(b) In a business venture a man can make a profit of £2000 with a probability of 0·65 or a loss of £4000 with a probability of 0·35. Find his expectation.

15. A stack of 20 cards contains four white, four green, four yellow, four red and four black. A sample of three cards is drawn at random from the stack. Find the probability that the three cards are: (a) All black, (b) All of one colour, (c) All of different colours.

16. How many four-digit numbers can be made from the digits 0, 2, 3, 5, 7, 9 if none of them appears more than once in each number? If each digit can appear as often as required how many numbers can be formed?

17. A and B play a game by alternately throwing two dice, A throwing first. The game is won by whoever first throws a score of ten or more. If A's total chance of winning is denoted by p express B's total chance of winning in terms of p (this can be done in two ways) and hence find p.

Verify your answer by considering the sum of A's chances of winning on the first, third, fifth, . . . throws.

18. A box contains ten radio valves all apparently sound although four of them are actually sub-standard. Find the chance that if two of the valves are taken from the box together they are both sub-standard.

19. When three marksmen take part in a shooting contest their chances of hitting the target are $\frac{1}{2}, \frac{1}{3}$, and $\frac{1}{4}$. Calculate the chance that one and only one missile will hit the target if all three men fire at it simultaneously.

20. An urn contains twelve balls, three of each of the colours red, green, blue and yellow. Three balls are drawn from the urn in succession without replacement. What is the joint probability that the first is red, the second green or blue, and the third yellow?

If this event has happened and two further balls are now drawn, what is the probability that at least one ball of each colour will have been drawn?

21. (a) A 'true' die is thrown; calculate the expectation of the score.

(b) Throw a die 60 times, noting the score. Find the mean of your results. Explain the difference between the value of the expectation obtained in (a) and the value of the mean obtained in (b).

22. If n people are allocated seats at random at a round table with n seats, what is the probability that two particular people are seated next to each other?

23. A box of 20 screws is known to contain five faulty ones. A sample of six is taken from the box. (a) In how many ways can the

sample be chosen? (b) How many of the possible samples will contain more than one faulty screw?

24. A box contains a very large number of components, 5 per cent of which are known to be defective. If a sample of ten is chosen at random what is the probability that the sample will contain one or more than one defective?

25. A certain local council consists of eight Socialists, two Liberals and five Conservatives. A committee of seven is randomly chosen from the council. Derive the probabilities that: (a) there will be only one Socialist, (b) there will be an odd number from each party, (c) each party will be represented on the committee. (Liv. U.)

26. (a) Five per cent of a large consignment of eggs are bad. Find the probability of getting at least one bad egg in a random sample of a dozen.

(b) A bag contains b black balls and w white balls, where b is greater than w. If they are drawn one by one from the bag, find the probability of drawing first a black, then a white and so on alternately until only black balls remain. (J.M.B.)

27. (a) A bag contains four red, four white and six black tokens. Three tokens are drawn at random. Calculate the probabilities that the three tokens are (i) two red and one black, (ii) exactly two of them are red.

(b) An amplifier circuit is made up of three valves. The probabilities, which are independent, of the three valves being defective are respectively $\frac{1}{20}, \frac{1}{25}$ and $\frac{1}{50}$. Calculate (i) the probability that the amplifier works, (ii) that the amplifier has one defective valve.

28. (a) A box contains three red, four white and five blue balls. Another box contains five red, six white and seven blue balls. One ball is drawn from each box. Find the probability that both balls are the same colour.

(b) A box contains six discs numbered 1–6. Find, for each integer k from 3 to 11, the probability that the numbers on two discs drawn without replacement have a sum equal to k.

(c) Four telephone numbers are chosen at random. Find the probability that their last digits are all different. (J.M.B.)

29. (a) A machine produces items some of which are defective. Each hour a sample of ten items is inspected, and if it contains no defectives the machine is allowed to run for another hour, otherwise the machine is stopped. Calculate (to two decimal places) the probability that the machine will not be stopped when the proportion of defectives is 10 per cent.

Find how large a sample should be inspected to ensure that if the proportion of defectives is 10 per cent, the probability that the machine will be stopped is at least 0·99.

(b) A man has five keys, only one of which will open a particular door. Find, for each integer k, the probability that the kth key will open the door if (i) the keys are tried in succession, (ii) the keys are tried at random.   (J.M.B.)

4

THE BINOMIAL AND POISSON DISTRIBUTIONS

4.1. THE BINOMIAL DISTRIBUTION

THE probability of a head when a certain coin is tossed is known to be p and the probability of a tail $q\,(=1-p)$. Consider, with their associated probabilities, all the possible results when three such coins are tossed.

Sequence	HHH	HHT	HTH	THH	HTT	THT	TTH	TTT
Probability	$p \times p \times p$ $= p^3$	$p \times p \times q$ $= p^2 q$	$p \times q \times p$ $= p^2 q$	$q \times p \times p$ $= p^2 q$	$p \times q \times q$ $= pq^2$	$q \times p \times q$ $= pq^2$	$q \times q \times p$ $= pq^2$	$q \times q \times q$ $= q^3$

All the above events are mutually exclusive (see section 3.2). Hence

$$\text{Pr(3H)} = p^3$$
$$\text{Pr(2H and 1T)} = \text{Pr(HHT or HTH or THH)} = p^2 q + p^2 q + p^2 q = 3p^2 q$$
$$\text{Pr(1H and 2T)} = \text{Pr(HTT or THT or TTH)} = pq^2 + pq^2 + pq^2 = 3pq^2$$
$$\text{Pr(3T)} = q^3$$

These form a complete set of mutually exclusive events and hence their probabilities add up to 1

$$q^3 + 3q^2 p + 3qp^2 + p^3 = 1 \qquad \qquad \dots\text{(4.1)}$$

The left-hand side of the equation (4.1) is the binomial expansion of $(q + p)^3$ and distributions of this type are called binomial distributions and are discrete probability distributions.

In general, if n coins are tossed then the probabilities associated with the events 0H's, 1H, 2H's, ... nH's, are given by the terms of the binomial expansion of $(q + p)^n$. Consider the $\text{Pr}[x\text{H's and }(n - x)T\text{'s}]$. For any particular arrangement of the H's and T's the probability is $p^x q^{n-x}$ (the events H on one coin and H or T on another are statistically independent, see section 3.3).

The number of arrangements of xH's and $(n - x)$ T's is the number of ways of choosing the x places for the heads (or the $n - x$ places for the tails) from n places, i.e. ^nC_x (or $^nC_{n-x}$ which is the same as ^nC_x (see equation 3.16), thus the $\text{Pr}(x\text{ H's and }(n - x)\text{ T's)} = {}^nC_x \, q^{n-x} p^x$ which is the $(x + 1)$th term of the expansion of $(q + p)^n$.

We thus have the following *binomial distribution*. If p is the probability of an event occurring during a trial (often referred to as the

59

probability of a success) and q is the probability of an event not occurring (probability of failure), then the probabilities associated with 0, 1, 2, ... n successes in n trials are given by the terms of the binomial expansion of $(q + p)^n$ thus:

No. of successes	0	1	2	...	$(n-1)$	n
Probability	q^n	$^nC_1q^{n-1}p$	$^nC_2q^{n-2}p^2$	...	$^nC_{n-1}qp^{n-1}$	p^n

$$\ldots\ldots (4.2)$$

Example 4.1. A man's chance of winning a game is $\dfrac{3}{5}$ If he plays five games what are the probabilities that he will win 0, 1, 2, ... 5 games respectively?

The probability of his winning a game is $\dfrac{3}{5}$, hence the required probabilities are given by the terms of the binomial expansion of

$$\left(\frac{2}{5}+\frac{3}{5}\right)^5 = \left(\frac{2}{5}\right)^5 + 5\left(\frac{2}{5}\right)^4\left(\frac{3}{5}\right) + 10\left(\frac{2}{5}\right)^3\left(\frac{3}{5}\right)^2 + \ldots + \left(\frac{3}{5}\right)^5$$

$$= \frac{1}{5^5}(32 + 240 + 720 + 1080 + 810 + 243)$$

| No. of successes | 0 | 1 | 2 | 3 | 4 | 5 |

Example 4.2. It is known that 5 per cent of a very large number of screws are oversize. What is the probability that in a sample of 20 screws there will be less than two oversize?

Since the number of screws is very large it can be assumed that the probability of obtaining an oversize screw is the same for all the 20 screws in the sample. Choosing any one of the screws can be regarded as an independent trial in which the probability of a 'success' (picking a defective) is $p = 0.05$, the number of trials is $n = 20$.

The required binomial expansion is $(0.95 + 0.05)^{20}$ ... (i). To find the probability of less than two oversize screws we add the probabilities of 0 and 1 oversize screws, that is the first and second terms in the expansion of expression (i).

$$\text{Required probability} = (0.95)^{20} + 20(0.95)^{19}(0.05)$$
$$= 0.3581 + 0.3770$$
$$= 0.7351$$

Example 4.3. Seven coins are tossed 256 times; calculate the expected frequencies for 0, 1, 2, ... 7 heads. Draw a histogram to represent the distribution (assume that the $Pr(H) = 0.5$ for all the coins).

Any single toss of the seven coins can be regarded as seven independent trials in which the probability of success (a head) is $p = 0.5$. Therefore the probabilities of 0, 1, 2, ... 7 heads are given by the terms of the binomial expansion:

$$(0.5 + 0.5)^7$$

Tables are available which give the terms of the binomial expansion, one is printed at the end of this book. From this (or by calculation) we obtain the probabilities detailed in column (ii) of *Table 4.1*.

Consider the probability of, say, 3 H and 4 T occurring in one throw of the seven coins, from the table this is 0.2734. Then in 256 throws we would expect 3 H and 4 H to occur $0.2734 \times 256 = 70$ times and hence column (iii) in *Table 4.1* gives the expected frequencies.

Table 4.1

Number of heads (i)	Probability (ii)	Expected frequency (iii)
0	0.0078	$0.0078 \times 256 = 2$
1	0.0547	$0.0547 \times 256 = 14$
2	0.1641	$0.1641 \times 256 = 42$
3	0.2734	$0.2734 \times 256 = 70$
4	0.2734	$0.2734 \times 256 = 70$
5	0.1641	$0.1641 \times 256 = 42$
6	0.0547	$0.0547 \times 256 = 14$
7	0.0078	$0.0078 \times 256 = 2$

It will be noted that the expected distribution for this binomial distribution is symmetrical. This is *not* so if $q \neq p$ (i.e. $p \neq 0.5$).

4.2. THE MEAN OF THE BINOMIAL DISTRIBUTION

The binomial distribution is a probability distribution of a variate x which takes the values 0, 1, 2, ... n, with probabilities $P_1, P_2, P_3, \ldots P_n$ given by the terms of the expansion of $(q + p)^n$. As we saw in section

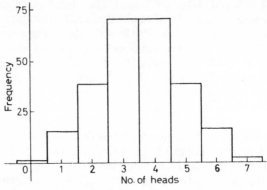

Figure 4.1. Histogram of expected (binomial) frequencies in 256 tosses of 7 coins

3.6 the mean or expected value of x is given by

$$\text{Mean} = \mathcal{E}(x) = \sum_{x=0}^{n} P_x x$$

Consider the expansion of

$$(q + pt)^n = q^n + {}^nC_1 q^{n-1} pt + {}^nC_2 q^{n-2} p^2 t^2 + \ldots + p^n t^n$$
$$= P_0 + P_1 t + P_2 t^2 + \ldots P_n t^n \qquad \ldots\text{(i)}$$

Differentiate both sides of equation (i) with respect to t, then

$$np(q + pt)^{n-1} = 0 + P_1 + 2P_2 t + \ldots nP_n t^{n-1} \qquad \ldots\text{(ii)}$$

Let $t = 1$ then, since $q + p = 1$, equation (ii) gives

$$np = 0 . P_0 + 1 . P_1 + 2 . P_2 + \ldots n . P_n,$$

i.e. $np = \sum\limits_{x=0}^{n} P_x x = \mathcal{E}(x) = \text{Mean} \qquad \ldots\text{(4.3)}$

Example 4.4. One hundred true coins were tossed and the number (x) of heads was noted. The experiment was repeated a large number of times. What is the expected number of heads?

Assuming that the coins are true we have that

$$p = 0\cdot5 = q, \text{ and } n = 100$$

$$\therefore \text{ by equation (4.3) } \mathcal{E}(x) = np$$
$$= 100 \times 0\cdot5$$
$$= 50$$

Exercises 4a

Where possible, check your answers by the tables at the back of the book.

1. Write down the terms of the following expansions: $(a) \left(\dfrac{1}{2} + \dfrac{1}{2}\right)^6$, $(b) \left(\dfrac{1}{4} + \dfrac{3}{4}\right)^5$, (c) $729\left(\dfrac{1}{3} + \dfrac{2}{3}\right)^6$, (d) $3125\left(\dfrac{1}{5} + \dfrac{4}{5}\right)^5$.

2. In a packet of flower seeds $\dfrac{2}{3}$ are known to give plants with yellow flowers and $\dfrac{1}{3}$ plants with red flowers.

(a) Calculate the probabilities of obtaining 0, 1, ... 5, 6 plants with yellow flowers in a row of six plants.

(b) If 729 rows of six plants are planted how many rows will be expected to have three yellow and three red flowering plants? [see question 1 (c)].

3. Assuming that the probability of any one of five telephone lines is engaged at any instant is $\dfrac{1}{4}$ calculate the probability that at any instant: (a) All five lines are engaged. (b) At least one of the lines is engaged.

4. If, on the average, rain falls on 18 days in every 30 days, find the probability that: (a) The first two days of a given week will be fine and the remainder wet. (b) Rain will fall on only three days of a given week.

5. In an industry workmen have a 25 per cent chance of suffering from an occupational disease. What is the probability that of seven workmen five or more will contract the disease?

6. The standard of efficiency of electric light bulbs is controlled by examining samples of five. *Table 4.2* shows the number of defective bulbs found in 100 samples.

Table 4.2

No. of defectives	0	1	2	3	4	5
No. of samples	63	27	8	1	1	0

Calculate the proportion of defective bulbs in the 500 bulbs tested. Assuming that the binomial distribution holds use this value to calculate the expected frequencies. (We have, as yet, no test for judging if the observed frequencies are statistically different from the expected frequencies. As we shall see later, the χ^2 test (Chapter 11) enables us to do this.)

7. By considering $\sum\limits_{x=0}^{n} P_x x$ where x has the values $0, 1, \ldots n$ and

$P_0 = q^n$, $P_1 = nq^{n-1}p$, $P_2 = \dfrac{n(n-1)}{1.2} q^{n-2}p^2 \ldots$ prove that the mean

of the binomial distribution is np. (*Hint*: remember that $q + p = 1$.)

4.3. THE POISSON DISTRIBUTION

The Poisson distribution is applicable in two different circumstances: (*a*) As an approximation to the binomial distribution. (*b*) When the variates occur in a random manner.

(*a*) When p (or q) is small and np (or nq) equal to a say, is finite, the Poisson distribution is a good approximation to the binomial distribution.

From the binomial distribution

$$\Pr(x \text{ successes in } n \text{ trials}) = {}^nC_x q^{n-x} p^x$$

If we let $n \to \infty$ and $p \to 0$ while np remains a fixed finite value a we

have that $np = a$, and $p = \dfrac{a}{n}$ .... (i)

$\Pr(x \text{ successes}) = {}^nC_x q^{n-x} p^x$

$$= \frac{\overbrace{n(n-1)(n-2)(n-3)\ldots(n-x+1)}^{x \text{ brackets}}}{x!} \cdot \frac{(1-p)^n}{(1-p)^x} \cdot p^x$$

(*Note*: $q = 1 - p$.)

$$= \frac{\overbrace{n(n-1)(n-2)\ldots(n-x+1)}^{x \text{ brackets}}}{x!} \cdot \frac{\left(1-\dfrac{a}{n}\right)^n}{\left(1-\dfrac{a}{n}\right)^x} \frac{a^x}{n^x} \quad \text{from (i)}$$

$$= \frac{1\left(1-\dfrac{1}{n}\right)\left(1-\dfrac{2}{n}\right)\ldots\left(1-\dfrac{x-1}{n}\right)}{x!} a^x \frac{\left(1-\dfrac{a}{n}\right)^n}{\left(1-\dfrac{a}{n}\right)^x}$$

As $n \to \infty$ $\left(1-\dfrac{a}{n}\right)^x$ (x finite) and $\left(1-\dfrac{1}{n}\right), \left(1-\dfrac{2}{n}\right) \ldots$ all tend to 1.

Also from the properties of the exponential function

$$\lim_{n \to \infty} \left(1 - \frac{a}{n}\right)^n = e^{-a}$$

Hence the $\Pr(x \text{ successes}) \to \frac{a^x}{x!} e^{-a}$ which is the Poisson probability.

The distribution is discrete and gives:

$$\Pr(0) = e^{-a}, \quad \Pr(1) = \frac{a}{1!} e^{-a}, \quad \Pr(2) = \frac{a^2}{2!} e^{-a},$$

$$\Pr(3) = \frac{a^3}{3!} e^{-a}, \qquad \ldots (4.4)$$

Note: the mean of the binomial distribution is $np (= a)$ and thus it is reasonable to assume that the mean of the Poisson distribution is a. It is proved equal to a in section 4.4.

Using the Poisson instead of the binomial distribution can reduce calculations quite considerably (see Example 4.5). The maximum values of p (or q) and the maximum value of np at which we should use the approximation depend on the accuracy required. As a rough guide p (or q) < 0.1 and $np < 6$ give a reasonable approximation.

Example 4.5. Oranges are packed in crates each containing 250. On the average 0·6 per cent are found to be bad when the crates are opened. What is the probability that there will be more than two bad oranges in a crate?

Here $p = 0.006$ and $n = 250$.

The binomial probabilities are given by the expansion of $(0.994 + 0.006)^{250}$ requiring calculations such as $(0.994)^{250}, {}^{250}C_1 (0.994)^{249} (0.006)$ etc.

However, since $np = 250 \times 0.006 = 1.5$

i.e. $a = 1.5$

we can make use of the Poisson distribution. In *Table 4.3* we have made use of a recurrence relation, since for the Poisson distribution

$$\Pr(x) = \frac{a^x}{x!} e^{-a}$$

$$= \frac{a}{x} \times \frac{a^{x-1}}{(x-1)!} e^{-a}$$

$$= \frac{a}{x} \Pr(x-1) \text{ we can simplify the calculations as shown (refer}$$

also to table at the back of the book).

Table 4.3

No. bad oranges	Probability (Poisson)		
0	$e^{-1\cdot5}$		0·2231
1	$\dfrac{1\cdot5}{1!}e^{-1\cdot5}$	$=\dfrac{1\cdot5}{1}\Pr(0)$	0·3347
2	$\dfrac{(1\cdot5)^2}{2!}e^{-1\cdot5}$	$=\dfrac{1\cdot5}{2}\Pr(1)$	0·2510

Total = 0·8088

Therefore Pr(> 2 bad oranges) $= 1 - 0\cdot8088$

$$= 0\cdot1912$$

(A practical answer in these circumstances would be that about one in five of the crates will contain more than two bad oranges.)

Example 4.6. In a table of random numbers (see section 2.5) the probability of occurrence of any given pair of digits is 0·01. In a set of 400 pairs of random numbers calculate, using both the binomial probabilities and the Poisson approximations, the probabilities of 0, 1, 2 or 3 pairs of sevens appearing in the set of 400 pairs.

The binomial probabilities are given by the terms of the expansion of $(0\cdot99 + 0\cdot01)^{400}$.

For the approximate values using Poisson probabilities we note that $a = np = 400 \times 0\cdot01 = 4$. Thus the approximate values are e^{-4}, $\dfrac{4}{1!}e^{-4}, \dfrac{4^2}{2!}e^{-4}, \dfrac{4^3}{3!}e^{-4}$.

Table 4.4

No. of pairs of sevens	Poisson probabilities	Binomial probabilities	
0	$e^{-4} = 0\cdot0183$	$(0\cdot99)^{400}$	$= 0\cdot0174$
1	$\dfrac{4}{1!}e^{-4} = 0\cdot0733$	$400(0\cdot99)^{399}(0\cdot01)$	$= 0\cdot0703$
2	$\dfrac{4^2}{2!}e^{-4} = 0\cdot1465$	$\dfrac{400.399}{1.2}(0\cdot99)^{398}(0\cdot01)^2$	$= 0\cdot1416$
	$\dfrac{4^3}{3!}e^{-4} = 0\cdot1954$	$\dfrac{400.399.398}{1.2.3}(0\cdot99)^{397}(0\cdot01)^3$	$= 0\cdot1898$

(b) Randomly occurring variates are to be found in the number of telephone calls in a certain time, particle emission, numbers of minute organisms in 1 ml of fluid, but there must be a random distribution. If the objects have a tendency to cluster, e.g. larvae eggs, then the Poisson distribution is not applicable.

As we have stated [see equations (4 4)], if a variate x has a Poisson distribution with an average or expected value of a, then the probabilities of $x = 0, 1, 2, \ldots r, \ldots$ are given by

$$e^{-a}, \frac{a}{1!}e^{-a}, \frac{a^2}{2!}e^{-a}, \ldots \frac{a^r}{r!}e^{-a}, \ldots$$

As this is a probability distribution the sum of the probabilities (to infinity) must be unity. This is easily verified

$$e^{-a} + \frac{a}{1!}e^{-a} + \frac{a^2}{2!}e^{-a} + \ldots = e^{-a}\left(1 + \frac{a}{1!} + \frac{a^2}{2!} + \ldots\right)$$

$$= e^{-a}e^{a}$$

$$= e^{0}$$

$$= 1$$

4.4. THE MEAN OF THE POISSON DISTRIBUTION

To find the mean of the distribution consider

$$e^{a(t-1)} = e^{-a}e^{at}$$

$$= e^{-a}\left(1 + at + \frac{a^2t^2}{2!} + \frac{a^3t^3}{3!} + \ldots\right)$$

Differentiate both sides with respect to t

$$a\,e^{a(t-1)} = e^{-a}\left(0 + a + 2.\frac{a^2t}{2!} + 3.\frac{a^3t^2}{3!} + \ldots\right) \qquad \ldots (i)$$

Let $t = 1$ in equation (i) and we have

$$a e^{0} = e^{-a}\left(0 + a + 2.\frac{a^2}{2!} + 3.\frac{a^3}{3!} + \ldots\right)$$

$$= ae^{-a} + 2\frac{a^2}{2!}e^{-a} + 3\frac{a^3}{3!}e^{-a} + \ldots$$

But $e^0 = 1$ and the Poisson probabilities $P_0, P_1, P_2, \ldots$ are e^{-a}, ae^{-a}, $\dfrac{a^2}{2!}e^{-a}, \ldots$ Hence

$$a = 0 \times P_0 + 1 \times P_1 + 2 \times P_2 + 3 \times P_3 \ldots$$

$$a = \sum_{x=0}^{\infty} xP_x$$

$$a = \mathscr{E}(x) = \text{Mean} \qquad \ldots (4.5)$$

Example 4.7. A manufacturer of resistors knows that 0·15 per cent of his output is outside the acceptable tolerance limits. The resistors are packed in crates each containing 2000 resistors. Find the probabilities of a crate containing 0, 1, 2, 3, ..., 6 or 7 resistors which are outside tolerance limits. What is the probability that a crate will contain more than seven defectives? Draw a histogram to represent the distribution.

The defective resistors are assumed to occur at random and this is therefore a Poisson distribution.

$$\text{Mean} = np = 2000 \times 0{\cdot}0015$$
$$= 3$$

The probabilities of 0, 1, 2, ... defectives are therefore e^{-3}, $\dfrac{3}{1}e^{-3}$, $\dfrac{3^2}{2!}e^{-3}$, ...The calculations are shown in *Table 4.5* (the answers could also have been obtained from the table at the back of the book).

Since the total of all the probabilities is one the probability of more than seven defectives is

$$\Pr(>7 \text{ defectives}) = 1 - 0{\cdot}9884$$
$$= 0{\cdot}0116$$

Example 4.8. A man has four cars for hire. The average demand on a weekday is for two cars. Assuming 312 weekdays per year obtain the theoretical frequency distribution of demands for cars during a weekday. Hence estimate, to the nearest whole number, the number of demands which would have to be refused in a year.

The demand is likely to be at random and we assume a Poisson distribution with 'a' = 2. The expected frequencies for different numbers of demands are given by 312 × probabilities and are as given in *Table 4.6*.

Table 4.5

No. defectives	Probability	
0	e^{-3}	0·0498
1	$\dfrac{3}{1!}e^{-3} = \dfrac{3}{1}\Pr(0)$	0·1494
2	$\dfrac{3^2}{2!}e^{-3} = \dfrac{3}{2}\Pr(1)$	0·2241
3	$\dfrac{3^3}{3!}e^{-3} = \dfrac{3}{3}\Pr(2)$	0·2241
4	$\dfrac{3^4}{4!}e^{-3} = \dfrac{3}{4}\Pr(3)$	0·1681
5	$\dfrac{3^5}{5!}e^{-3} = \dfrac{3}{5}\Pr(4)$	0·1009
6	$\dfrac{3^6}{6!}e^{-3} = \dfrac{3}{6}\Pr(5)$	0·0504
7	$\dfrac{3^7}{7!}e^{-3} = \dfrac{3}{7}\Pr(6)$	0·0216

Total = 0·9884

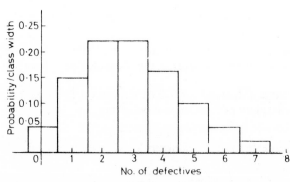

Figure 4.2. Histogram of the expected (Poisson) probabilities of the distribution in Table 4.5.

Table 4.6

No. of demands	Expected frequency	
0	$312 \times e^{-2} \qquad = 312 \times 0.1353$	42.213
1	$312 \times \dfrac{2}{1!}e^{-2} \quad = \dfrac{2}{1}\text{Frequ(0)}$	84.43
2	$312 \times \dfrac{2^2}{2!}e^{-2} \quad = \dfrac{2}{2}\text{Frequ(1)}$	84.43
3	$312 \times \dfrac{2^3}{3!}e^{-2} \quad = \dfrac{2}{3}\text{Frequ(2)}$	56.28
4	$312 \times \dfrac{2^4}{4!}e^{-2} \quad = \dfrac{2}{4}\text{Frequ(3)}$	28.14
5	$312 \times \dfrac{2^5}{5!}e^{-2} \quad = \dfrac{2}{5}\text{Frequ(4)}$	11.26
6	$312 \times \dfrac{2^6}{6!}e^{-2} \quad = \dfrac{2}{6}\text{Frequ(5)}$	3.75
7	$312 \times \dfrac{2^7}{7!}e^{-2} \quad = \dfrac{2}{7}\text{Frequ(6)}$	1.07
8	$312 \times \dfrac{2^8}{8!}e^{-2} \quad = \dfrac{2}{8}\text{Frequ(7)}$	0.27
9	$312 \times \dfrac{2^9}{9!}e^{-2} \quad = \dfrac{2}{9}\text{Frequ(8)}$	0.06
10	$312 \times \dfrac{2^{10}}{10!}e^{-2} = \dfrac{2}{10}\text{Frequ(9)}$	0.01

In calculating the number of demands that have to be refused we have to remember that whatever the demand only four cars are available.

From the table a demand for five cars arises on 11·26 days and one demand is refused on each of these days giving:

$$1 \times 11.26 \text{ demands refused.}$$

Similarly from the table a demand for six cars arises on 3·75 days giving

$$2 \times 3.75 \text{ demands refused.}$$

Proceeding in this way the number of demands refused is:

$11.26 \times (5 - 4) + 3.75 \times (6 - 4) + 1.07 \times (7 - 4) + \ldots$

$= 11.26 + 7.50 + 3.21 + 1.08 + 0.29$

$= 23$ (to the nearest whole number) demands refused per year.

4.5. THE ADDITIVE PROPERTY OF THE POISSON DISTRIBUTION

The combination of two Poisson distributions with means a and b is itself a Poisson distribution with mean $a + b$ (the events giving rise to the two distributions being independent).

Proof

For the first source

$$\Pr(0) = e^{-a}, \Pr(1) = ae^{-a}, \Pr(2) = \frac{a^2}{2!}e^{-a}, \ldots$$

For the second source

$$\Pr(0) = e^{-b}, \Pr(1) = be^{-b}, \Pr(2) = \frac{b^2}{2!}e^{-b}, \ldots$$

For the combined sources

$$\Pr(0) = \Pr(0)\Pr(0) \qquad = e^{-a}e^{-b} \qquad = e^{-(a+b)}$$

$$\Pr(1) = \Pr(0)\Pr(1) + \Pr(1)\Pr(0) = e^{-a} . be^{-b} + ae^{-a} . e^{-b}$$
$$= (a + b)e^{-(a+b)}$$

$$\Pr(2) = \Pr(0)\Pr(2) + \Pr(1)\Pr(1) + \Pr(2)\Pr(0)$$
$$= e^{-a}\frac{b^2}{2!}e^{-b} + ae^{-a} . be^{-b} + \frac{a^2}{2!}e^{-a} . e^{-b} \qquad = \frac{(a + b)^2}{2!}e^{-(a+b)}$$

Similarly

$$\Pr(3) \qquad\qquad\qquad\qquad\qquad = \frac{(a + b)^3}{3!}e^{-(a+b)}$$

$$\Pr(4) \qquad\qquad\qquad\qquad\qquad = \frac{(a + b)^4}{4!}e^{-(a+b)}$$

. . . .

The right-hand column contains the probabilities for a Poisson distribution with mean $(a + b)$.

D

Exercises 4b

1. A manufacturer packs his components in boxes containing 100 each. In general 3·5 per cent of the components are known to be defective. Find the probability that a box contains more than six defectives.

2. It is known that 0·1 per cent of electric light bulbs are defective. What is the probability that in a crate containing 1000 bulbs there are more than three defective bulbs? What is the probability that there are more than four defectives in two crates each containing 1000 bulbs?

3. Calculate, correct to three significant figures, the first five terms of the binomial expansion of $\left(\dfrac{35}{36} + \dfrac{1}{36}\right)^{36}$. From the tables, or by calculation, obtain the first five terms of the Poisson approximation. Put your results into tabular form so that the two sets may be compared.

4. A distributor of seeds knows from large-scale tests that 2 per cent of wallflower seeds will not germinate. He sells the seeds in packets of 200 and guarantees 95 per cent germination. What percentage of packets can be expected to violate the guarantee?

5. Calculate the mean of the distribution in *Table 4.7*. Assuming that the Poisson law holds calculate the expected frequencies using the calculated mean.

Table 4.7

No. of defects per machine	0	1	2	3	4	5
Frequency of machines	15	30	23	18	10	4

EXERCISES 4

(Wherever possible check your answers by the tables at the back of the book.)

1. Find the probability that in a family of six children there will be: (*a*) At least one boy. (*b*) At least one boy *and* one girl. (*c*) Three boys and three girls.
(Assume that the probability of a male birth is 0·5).

2. If 20 per cent of the packages produced by an automatic machine are defective find the probability that if five packages are chosen at random: (*a*) one is defective, (*b*) three are defective, (*c*) at most three are defective.

3. Calculate the first four terms of the Poisson distribution when the mean is (*a*) 5, (*b*) 0·5, (*c*) 3·2.

4. A car hire firm has two cars which it hires out by the day. The daily demand for cars follows a Poisson distribution with mean 1·3. If both cars are used equally, calculate the probabilities of: (a) No car being hired. (b) A particular car *not* being hired. (c) Both cars being hired on any one day.

5. The number of direct hits on dwelling houses during flying bomb raids in 100 small areas of London are given in *Table 4.8*.

(a) Give reasons why this may be expected to follow a Poisson distribution.

(b) Calculate the mean of the distribution.

(c) Calculate the expected frequencies assuming a Poisson distribution (give your answers correct to one decimal place).

Table 4.8

No. of hits	0	1	2	3	4	5
No. of areas	24	35	24	12	4	1

6. The probability that a student entering university will graduate is 0·9. Determine the probability that of four students: (a) None will graduate. (b) At least one will graduate. (c) Not more than two will fail.

7. If 3 per cent of electric light bulbs are defective find the probabilities that in a batch of 100 bulbs: (a) 0, 1, or 2 bulbs are defective. (b) More than five are defective.

8. In a large consignment of electric light bulbs 5 per cent of the bulbs are defective. Calculate the probability that a random sample of 20 will contain at most two defective bulbs. Of a number of consignments which are to be inspected one third have 5 per cent of their total bulbs defective, the rest have 10 per cent defective. If a consignment is rejected when a random sample of 20 taken from it contains more than two defective bulbs, find the proportion of the consignments which are rejected.

9. The average number of telephone calls per hour is 30. Calculate the probabilities of 0, 1, 2, 3, or 4 calls being made in any period of 2 minutes.

10. The average number of cars crossing a narrow bridge is 150 per hour. If four or more cars attempt to cross in any one minute a queue forms. If the traffic is flowing freely what is the probability of a queue forming?

11. Of a large mass of components 12·5 per cent are known to be defective. A sample of eight is picked at random. Using (a) the binomial distribution, (b) the Poisson distribution find the probability that exactly three defectives are chosen.

12. The average monthly demand for a certain article from store is three. During a two-year period of limited supplies it is only possible to clear off outstanding demands and to make the total stock up to four at the beginning of each month. Estimate, using a Poisson distribution, the total number of demands refused during the two years.

13. Calculate the first four terms of the binomial distribution $(q + p)^n$ with: (a) $p = \dfrac{1}{2}$, $n = 8$, (b) $p = \dfrac{1}{5}$, $n = 20$, (c) $p = 0 \cdot 1$, $n = 40$. (d) Also of the Poisson distribution with $a = 4$. Put your results in a tabular form to illustrate how the binomial approaches a Poisson distribution as n becomes larger and np remains finite.

14. Twenty coins are taken at random from a box. The box contains pennies and halfpennies mixed together in the ratio of three pennies to two halfpennies. Find the probability that the twenty coins:

(a) Will contain exactly 15 pennies.

(b) Can be shared equally between five children so that they each receive the same number of pennies, this number being at least two.

15. In a large quantity of mixed flower seed the chance of a seed giving rise to a plant with blue flowers is one in seven. Show that the chance of eight or more plants with blue flowers from any ten seeds is about 1 in 168,000.

The seeds are sown and the resulting plants put into groups of ten and the number of cases of blue flowers in each group is recorded. Estimate the mean of these numbers correct to three significant figures (assume that all the seeds have an equal chance of germination).

16. The numbers below are the frequencies with which telephone calls are received at a business office in 17 successive half-hour periods throughout the day:

2, 2, 1, 3, 2, 3, 4, 3, 2, 0, 3, 4, 1, 3, 2, 3, 2.

Give reasons for supposing that these observations might follow a Poisson distribution. Calculate the theoretical frequencies assuming that they do follow a Poisson distribution.

17. The number of α-particles radiated from a disc in 2608 periods of 7·5 sec were counted by Rutherford and Geiger.

Table 4.9

No. of α-particles	0	1	2	3	4	5	6	7	8	9	10	11	12
Frequency of occurrence	57	203	383	525	532	408	273	139	45	27	10	4	2

Fit a Poisson distribution to this data by calculating the expected frequencies. (Liv. U.)

18. Sweets are sold in packets of five. *Table 4.10* shows the distribution of purple sweets in 100 packets.

Table 4.10

No. of purple sweets	0	1	2	3	4	5
No. of packets	73	16	6	3	1	1

Calculate the mean number of purple sweets per packet and assuming that the binomial distribution applies, estimate the probability of obtaining a purple sweet when one is chosen at random from a packet.

19. The number of people using or requiring use of the telephone service of the university at any given instant has a Poisson distribution with a mean of three. If the number of lines in the switchboard is five determine the probability that all the lines are required at a given instant. If a call is directed to any one of the available lines at random find the probability that a particular line is not required at a given instant.

It is desired that the probability of all the lines being required at a given instant is not greater than 0·1. Suggest how this may be realized.
(Liv. U.)

20. Prove that, if $n \to \infty$ and $p \to 0$ such that $np = m$

$$\binom{n}{k} p^k (1 - p)^{n-k} \to \exp(-m) \frac{m^k}{k!}$$

Past experience in the production of a certain component has shown that the proportion of defectives is 0·003. Components leave the factory in boxes of 500. What is the probability that (a) a box contains three or more defectives, (b) two successive boxes contain six or more defectives between them? $\left[\binom{n}{k} = {}^nC_k \right]$ (Liv. U.)

21. A variate x takes the values 0, 1, 2, 3, 4, 5 with probabilities equal respectively to the terms, arranged in the usual way, of the binomial expansion of $(p + q)^5$, where $p + q = 1$. Show that the mean value of x is $5q$.

Razor blades of a certain kind are sold in packets of five. *Table 4.11* shows the frequency distribution of 100 packets according to the number of faulty blades contained in them:

Table 4.11

No. of faulty blades	0	1	2	3	4	5
No. of packets	84	10	3	2	1	0

Calculate the mean number of faulty blades per packet and, assuming that the binomial law applies, estimate the probability that a blade taken at random from any packet will be faulty.   (J.M.B.)

5

MEASURES OF CENTRAL TENDENCY

5.1. INTRODUCTION

IN Chapter 2 we showed how to reduce a mass of data to a more understandable form by classifying or grouping it and then depicting it diagrammatically. In this and the next chapter we shall consider certain statistical parameters which measure the principal characteristics of a distribution. The first parameter is a measure of the centre of a distribution. As we have seen in section 3.6 the mean (arithmetic mean) μ of a distribution is equal to the expectation of x. Also the sample mean $\bar{x}$ is an estimate of μ so that it is natural to use the mean or the sample mean as an estimate of central tendency.

The following examples illustrate the way in which the mean may be calculated.

5.2. THE MEAN

Referring back to equation (3.20) we have that

$$\text{Mean} = \frac{\sum_{i=1}^{n} f_i x_i}{\sum_{i=1}^{n} f_i} \qquad \dots (5.1)$$

Example 5.1. Find the mean number of potatoes per plant given the following frequencies of occurrence:

No. of potatoes per plant (x)	2	3	4	5	6	7	8	9
No. of plants (f)	4	5	10	29	30	18	3	1

$$\bar{x} = \frac{4 \times 2 + 5 \times 3 + 10 \times 4 + 29 \times 5 + 30 \times 6 + 18 \times 7 + 3 \times 8 + 1 \times 9}{4 + 5 + 10 + 29 + 30 + 18 + 3 + 1}$$

$$= \frac{547}{100}$$

$$= 5{\cdot}47$$

If the data has been grouped into classes we allocate the mid-interval value (see section 2.2) to each member of the class.

Example 5.2. Find the mean of the following distribution:

x	35–40	–45	–50	–55	–60	–65
f	7	18	23	24	16	12

The mid-points of the classes are 37·5, 42·5, 47·5, 52·5, 57·5 and 62·5, thus the mean is given by

$$\frac{1}{100}(7 \times 37{\cdot}5 + 18 \times 42{\cdot}5 + 23 \times 47{\cdot}5 + 24 \times 52{\cdot}5 + 16 \times 57{\cdot}5$$

$$+ 12 \times 62{\cdot}5)$$

$$= \frac{1}{100}(5050{\cdot}0)$$

$$= 50{\cdot}5$$

The calculations can generally be shortened by use of a *false origin*, e.g. in finding the mean of the numbers 997, 999, 1001, 1004, and 1006 we can take a false origin of 1000 and subtract it from all the numbers this gives, $-3, -1, +1, +4, +6$. The average of these is $\frac{1}{5}(-3 -1 +1 +4 +6) = \frac{7}{5}$.

This is the amount by which the average differs from 1000 and hence the average is $1000 + \frac{7}{5} = 1001{\cdot}4$.

The choice of 1000 is purely arbitrary and was chosen because it is near to the centre of the values and gave small differences.

Now consider the numbers (x_i) 725, 775, 800, 850, 875, 950. A suitable false origin is 800 and the differences (deviations) are $(x_i - 800)$ -75, $-25, 0, 50, 75, 150$.

In this case a further simplification is possible by removing the common factor 25 giving $\left(\dfrac{x_i - 800}{25}\right)$, i.e.

$$-3, -1, 0, 2, 3, 6.$$

The average of these is

$$\frac{1}{6}(-3 - 1 + 0 + 2 + 3 + 6) = \frac{7}{6}.$$

However, we are dealing with units of 25, thus the mean is:

$$800 + 25 \times \frac{7}{6} = 829\frac{1}{6}$$

In general, if a is the false origin, C the common factor, and x the variable, then the deviation d is given by

$$d = \frac{x - a}{C} \qquad \qquad \dots (5.2)$$

Rewriting, we have $x = a + Cd$.

Now $$\bar{x} = \frac{\Sigma fx}{\Sigma f}$$

and substituting for x we have

$$\bar{x} = \frac{\Sigma f(a + Cd)}{\Sigma f}$$

$$= \frac{\Sigma af + \Sigma fCd}{\Sigma f}$$

C and a are constants

$$\therefore \qquad \bar{x} = a\frac{\Sigma f}{\Sigma f} + C\frac{\Sigma fd}{\Sigma f}$$

$$= a + C \frac{\Sigma fd}{\Sigma f} \qquad \qquad \dots (5.3)$$

The use of this formula can simplify the arithmetical calculations as illustrated in the following example.

Example 5.3. Find the arithmetical mean of the distribution given in columns (*i*) and (*ii*) of *Table 5.1*.
The calculations are also shown in the Table.
(*a*) The mid-points (X) of the classes are shown in column (*iii*) and 2·45 has been chosen as a false origin ($a = 2\cdot45$).
(*b*) In column (*iv*) the values (X) in column (*iii*) have been referred to this false origin ($X - 2\cdot45$).
(*c*) In column (*v*) the values in column (*iv*) have been divided by the common factor 0·1 $\left(d = \dfrac{X - 2\cdot45}{0\cdot1}\right)$.

Table 5.1

(i) (x)	(ii) (f)	(iii) Mid-points of classes (X)	(iv) X − 2·45	(v) $d = \dfrac{X - 2 \cdot 45}{0 \cdot 1}$	(vi) (fd)	(vii) (d + 1)	(viii) f(d + 1)
2·00 up to 2·10	16	2·05	−0·40	−4	−64	−3	−48
up to 2·20	85	2·15	−0·30	−3	−255	−2	−170
up to 2·30	200	2·25	−0·20	−2	−400	−1	−200
up to 2·40	312	2·35	−0·10	−1	−312	0	−418
					−1031		
up to 2·50	206	2·45	0	0		1	206
up to 2·60	112	2·55	0·10	1	112	2	224
up to 2·70	95	2·65	0·20	2	190	3	285
up to 2·80	62	2·75	0·30	3	186	4	248
up to 2·90	20	2·85	0·40	4	80	5	100
up to 3·00	5	2·95	0·50	5	25	6	30
					593		1093

Totals $\Sigma f = 1113$ $\Sigma fd = -438$ $\Sigma f(d + 1) = 675$

(d) In **column** (vi) the products of corresponding values of column (ii) and column (v) are summed (advantage being taken of the zero opposite $X = 2\cdot45$ to sum the negative values separately) and Σfd obtained.

(e) Application of formula (5.3) gives $\bar{x}$.

(f) Columns (vii) and (viii) provide a very useful check and should always be completed. $\Sigma f(d + 1) - \Sigma fd = \Sigma f$, that is the sum of column (viii) minus the sum of column (vi) must equal the sum of column (ii).

$$\text{Check } \Sigma f(d + 1) - \Sigma fd = 675 - (-438)$$

$$= 1113$$

$$= \Sigma f$$

Applying the formula (5.3)

$$\bar{x} = 2\cdot45 + 0\cdot10 \left(\frac{-438}{1113}\right)$$

$$= 2\cdot45 - 0\cdot039$$

$$= 2\cdot41 \text{ (to 3 significant figures)}$$

When using desk calculating machines a false origin does not simplify the calculations a great deal.

Exercises 5a

1. Calculate the mean value for the samples given in *Table 5.2*.

Table 5.2

(a)		(b)	
Length of screws to nearest 0·01 in	(f)	Daily wage (shillings)	(f)
3·80–3·89	3	20–30	3
3·90–3·99	5	–40	53
4·00–4·09	14	–50	72
4·10–4·19	19	–60	28
4·20–4·29	28	–70	19
4·30–4·39	18	–80	12
4·40–4·49	10	–90	6
4·50–4·59	3	–100	4
Total	100	–110	3
		Total	200

2. Calculate the mean value of distribution in *Table 5.3* which is concerned with the age distribution of children. (Note the unequal class widths and that the ages are not rounded off.)

Table 5.3

Age (years)	Frequency (1000s)
0–3 incl.	44
4–5 incl.	29
6–11 incl.	98
12	17
13	16
14	15
15	16
16	4
17–19 incl.	13
20	1
Total	253

3. Find the mean of the distributions in *Table 5.4*.

4. In a study of the yields of two types of peas, the number of peas in 1000 pods of each variety was noted and is given in *Table 5.5*.

Find the mean of each distribution.

5. An automatic machine is set to give a screw of length 2·50 in. A sample of 100 screws is taken and each one measured to the nearest 0·01 in. The distribution of lengths being given in *Table 5.6*.

A difference of more than 0·015 in between the sample mean and the supposed setting of the machine and the machine has to be reset. Is there any reason to reset the machine?

6. The mean weight of 12 packets is 16·1 oz; another 17 packets have a mean weight of 15·8 oz; what is the mean weight of all the 29 packets together?

In general, the sample mean is used as the estimate of the central tendency. Other estimates are, however, available and in sections 5.3, 5.4 and 5.5 we shall briefly discuss the *median, mode* and *geometric mean.*

Table 5.4

x	(a) f	(b) f
10–20	78	30
–30	160	71
–40	121	89
–50	85	52
–60	63	31
–70	21	15
–100	11	10
Totals	539	298

Table 5.5

No. of peas in a pod	0	1	2	3	4	5	6	7	8
Type A	5	50	87	100	105	190	272	105	86
Type B	6	24	38	72	176	209	205	150	120

Table 5.6

Length	2·47	2·48	2·49	2·50	2·51	2·52	2·53	2·54
Frequency	2	6	9	13	23	19	20	8

5.3 THE MEDIAN

Given a set of discrete variates arranged in order of magnitude then the *median* is the value of the variate which divides the set into two numerically equal groups. If there are an even number of members in the set the median is taken as half the sum of the values of the variates ascribed to the two middle members.

Example 5.4. The median value of the set:

$$2{\cdot}0, 2{\cdot}0, 2{\cdot}1, 2{\cdot}2, 2{\cdot}2, 2{\cdot}3. \text{ is } \frac{2{\cdot}1 + 2{\cdot}2}{2}, \text{ that is } 2{\cdot}15.$$

If the variates have been arranged in groups the median is the value of the variate which divides the histogram into equal areas.

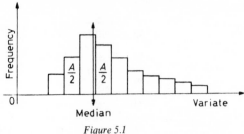

Figure 5.1

In the case of grouped variates the table of values can only indicate the *median class* (i.e. the class in which the median lies). To find the median we use the cumulative frequency curve (see section 2.3) as in the following example.

Example 5.5. Table 5.7 gives the wage distribution of part-time and full-time factory employees. Find the median wage.

Table 5.7

Wage (£'s per annum)	Frequency	Cumulative frequency
0–200	40	40
201–400	20	60
401–500	128	188
501–600	180	368
601–700	225	593
701–800	120	713
801–1000	43	756
1001–1500	17	773
1500–2000	8	781

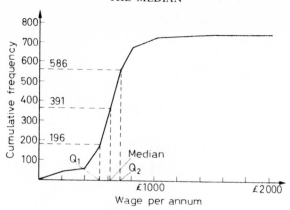

Figure 5.2. To illustrate the data of Table 5.7

Because the total population is 781 we require the value of the variate corresponding to the value 391 and from the graph this can be found to be £610.

The median is often known as the 50 *percentile*. Percentiles are the values of the variate dividing the distribution into hundredths. Other often used percentiles are the 25 and 75 percentiles. These are known as the lower and upper *quartiles* (Q_1 and Q_2) for obvious reasons. Five percentiles are often used by examining boards.

Since the points of the cumulative frequency graph are joined by straight lines it is possible to calculate the percentiles by *linear interpolation*. To calculate the median consider the part of the cumulative frequency graph (*Figure 5.2*) which is shown enlarged in *Figure 5.3*.

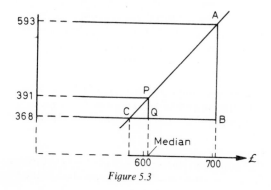

Figure 5.3

By similar triangles $\dfrac{PQ}{AB} = \dfrac{CQ}{CB}$

i.e. $\dfrac{391 - 368}{593 - 368} = \dfrac{CQ}{700 - 600}$

$\therefore$ $CQ = \dfrac{391 - 368}{593 - 368} \times £100$

$= £10\cdot2$

$\therefore$ Median $= £600 + £10\cdot2$

$= £610$ (a value of £610 is as accurate as can be expected from the original data)

5.4. THE MODE

The *mode* of a distribution is the value of the variate with the largest frequency. In the case of a classified distribution the class for which the frequency is greatest is called the *modal class* and the point which divides the modal class width in the inverse ratio of the two adjacent frequencies is known as the *crude mode*. There may be more than one modal class.

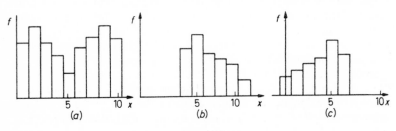

Figure 5.4

The modal class is easy to find either from a table of classified data or from a diagram of the distribution. It can, however, be an unreliable measure of the centre of a distribution [refer to the two dissimilar distributions shown in *Figures 5.4(b)* and (c), which have the same modal class].

It is very seldom used but can, occasionally, be of value when applied, for example, to the output of workers under various conditions. The modal conditions are those most conducive to maximum efficiency.

Exercises 5b

1. *Table 5.8* shows the distribution of the life (in hours) of 200 electric light bulbs. Find the crude mode and the median values.

Table 5.8

Life (h)	590–600	up to 610	up to 620	up to 630
Frequency	5	8	21	43
Life (h)	up to 640	up to 650	up to 660	up to 670
Frequency	78	29	12	4

2. If in *Table 5.8* subsequent checking of the data reveals that a constant error has been made in all the observations, find, where possible, the true value of the crude mode and median when the constant error is such that each value of the life of a bulb was:

(a) 20 h too high.

(b) Half what it should be.

(c) One-and-a-half times what it should be.

3. In the table in question 1, if the values of the frequencies had been written in reverse order what would be the value of the crude mode and the median?

4. The lengths of 100 steel rods are measured to the nearest 0·01 in and the results are as given in *Table 5.9*.

Table 5.9

Length to the nearest 0·01 in	Frequency
3·80–3·89	3
3·90–3·99	5
4·00–4·09	14
4·10–4·19	19
4·20–4·29	28
4·30–4·39	18
4·40–4·49	10
4·50–4·59	3
	Total 100

Find the value of the crude mode and the median.

5. Draw a cumulative frequency curve for the data in *Table 5.10* and hence obtain estimates of the median and the quartiles of the distribution.

Table 5.10

Daily wage (shillings)	No. of wage earners
20–30	3
–40	53
–50	72
–60	28
–70	19
–80	12
–90	6
–100	4
–110	3
Total	200

The mean is often referred to as the average; unfortunately 'average' is also applied to the median or the mode and hence our readers must be careful to decide which average is being used. For example, consider the following:

Example 5.6. In a small hamlet consisting of 10 cottages, 1 shop and 1 large house the weekly income of each of the tenants of the cottages is £12, of the shopkeeper £15 and of the owner of the large house £165. Find the median and the mean weekly wage of the population. Discuss their uses as the 'average' wage of the hamlet.

$$\text{Arithmetic mean} = \frac{1}{12}(10 \times £12 + £15 + £165)$$

$$= \frac{1}{12} £300$$

$$= £25 \text{ per week}$$

$$\text{Median} = £12 \text{ per week}$$

In this extreme case the arithmetic mean gives a false picture as it is bigger than the incomes of 11 out of the 12 inhabitants. The median gives a truer picture. This is a very skew distribution (as wage patterns often are) and the median is, in this case, a better 'average'. However, in a large number of practical applications we have distributions which are only slightly skew and the mean is to be preferred as a measure of central tendency.

5.5. THE GEOMETRIC MEAN

The *geometric mean* of *n* numbers *a*, *b*, *c*, ... is the *n*th root of their product.

$$\text{Geometric Mean (G.M.)} = \sqrt[n]{(abc \dots)} \qquad \dots (5.4)$$

Thus the G.M. of the four numbers 10, 24, 27, 125 is

$$\sqrt[4]{(10 \times 24 \times 27 \times 125)} = 30$$

The geometric mean is not often used but is sometimes useful in economic statistics where percentage rises (or falls) are often used. For example, the information that the price of an article has risen by 1s. 0d. is not enough, we need to know the original cost of the article. If its original cost was 2s. 0d. this is a 50 per cent rise. If its original cost was 100s. 0d. this is only a 1 per cent rise. A steady rise of 5 per cent per annum would mean that the actual prices (say 1000, 1050, 1102·5 ...) form a geometric progression. It is in these cases where the data are (or approximate to) a geometric progression that a geometric mean is used.

Example 5.7. The following index numbers indicate the price of an article as a percentage of its price in the year 1950. Find their geometric mean.

Year	1950	1955	1960	1965
Index	100	118	147	178

Index number	Logarithm
100	2·0000
118	2·0719
147	2·1673
178	2·2504

4) 8·4896

Log. of the Index = 2·1224

Index = 133 (to the nearest whole number)

EXERCISES 5

1. For a selected group of people the records of an insurance company gave the data in *Table 5.11*.

Table 5.11

Age	0–5	–10	–15	–20	–25	–30	–35	–40	–45
No. of deaths	3	0	5	16	19	82	95	81	45
Age	–50	–55	–60	–65	–70	–75	–80	–85	above 85
No. of deaths	50	28	18	10	6	3	2	1	1

Find the median and the mean of the distribution correct to two decimal places.

2. The average value of one sample of 24 resistors is 102 Ω and the average value of another sample of 36 resistors is 97 Ω, what is the grand average of the 60 resistors?

3. Two distributions have total frequencies f_1 and f_2 and their arithmetic means are $\bar{x}_1$ and $\bar{x}_2$ respectively. What is the arithmetic mean of the distribution obtained by combining the two?

4. In 1959 the age distribution of the population of the United Kingdom was as follows.

Age (years)	0–20	–30	–40	–50	–60	–80	above 80
Number (100,000s)	156	66	74	69	68	77	10

Decide on a reasonable limiting value for the last group and draw a histogram of the distribution. Also find the median and the mean.

5. In the following frequency distribution:

x	10–15	–20	–25	–30	–35	–40
f	12	16	17	23	29	28

considering the values of the variables to be at the mid-points of each class, find the value of $\Sigma f(x - 27{\cdot}5)$. What can be deduced from the result?

6. Two thousand candidates were given a test, the marks for which were zero or one of the integers 1–20. *Table 5.12* gives the number of candidates obtaining each mark:

Table 5.12

Mark	0	1	2	3	4	5	6	7	8	9
Frequency	6	39	43	82	124	145	152	187	191	226
Mark	10	11	12	13	14	15	16	17	18	19	20
Frequency	165	187	145	89	84	63	38	14	10	6	4

Find the mean mode and median of this distribution.

7. *Table 5.13* gives the marks of 310 candidates in an examination. Thus there is one candidate with 6 marks, three candidates with 23 marks and so on:

Table 5.13

Tens	0	1	2	3	Digits 4	5	6	7	8	9	Totals
0							1				1
10						1	1	1		1	4
20	1	2	1	3	1	4	1	1	1	4	19
30	10	8	8	9	16	13	8	3	11	4	90
40	9	5	2	8	4	10	6	2	8	10	64
50	17	9	3	3	4	7	4	8	6		61
60	11	10	2	5	3	8	5	1	3	8	56
70	5	2	1		2	3	1				14
80	1										1
	54	36	17	28	30	46	27	16	29	27	310

Calculate the mean average mark for the whole group. Estimate the mean from *Table 5.14*.
To what do you attribute the difference between these results?

Table 5.14

Range of marks	0–9	10–19	20–29	30–39	40–49	50–59	60–69	70–79	80–100
No. of candidates	1	4	19	90	64	61	56	14	1

8. Seventy screws produced by an automatic machine have lengths (measured to the nearest 0·01 cm) as given in *Table 5.15*.
Illustrate the table by a histogram and comment on its shape. Calculate the arithmetic mean and mark it on your graph.

Table 5.15

Length (cm)	4·40–4·49	4·50–4·59	4·60–4·69	4·70–4·79	4·80–4·89	4·90–4·99
No. of screws	5	8	14	20	17	6

9. Find the geometric mean of the numbers: 2·916, 3·888, 5·184, 6·912, 9·216, 12·288, 16·384.

10. A distribution has values of the variate x_1, x_2, ... x_n with corresponding frequencies f_1, f_2, ... f_n. A new distribution with the same frequencies is formed by taking $X_r = 2x_r - 3$ for all values of r ($= 1$, 2, ... n). If the values of the mean median and mode of the original distribution are a, b, and c respectively, what are their values for the new distribution?

11. A salesman makes a trip by car lasting 6 days. Each day he covers 220 miles. His average speeds for the 6 days are (in m.p.h.) 48, 36, 42, 48, 30, 36. What is his average speed for the whole trip?

12. For a moderately skew distribution, mean − mode ≏ 3(mean − median). Verify that this relation is approximately true for the data of question 6.

13. The average age of a group of 100 men is 30·8 years. If 38 of the men are over 35 and their average age is 45 years, what is the average age of the men under 35?

14. Show that the arithmetic mean of an arithmetic progression is the mean of the first and last terms, and that the geometric mean of a geometric progression with an odd number of terms is the middle term.

6

MEASURES OF DISPERSION

6.1. INTRODUCTION

IN the previous chapter we considered the statistical parameters which could be used as a measure of the centre of a distribution, the most important being the mean. In this chapter are discussed some parameters which can be used as a measure of the spread of a distribution. These are the variance, the mean deviation and the range. Of these the most important is the variance.

Consider the following histograms which represent the distribution of weights delivered by two automatic packing machines (*Figure 6.1*).

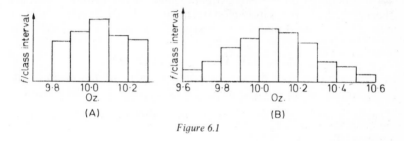

Figure 6.1

Both distributions have the same mean but whilst one is grouped close to the mean the other is more spread out. That is, machine A is giving a smaller variation in the delivered weights than machine B. The manufacturer using machine A would be able to guarantee that a much higher percentage of his product (mean weight 10·05) was within a certain distance of the mean than would the manufacturer using machine B.

Also it can be seen [see *Figure 6.2*] that if we adjust the packing machines so that 95 per cent of the packages delivered are greater in weight than 16 oz, then machine A can be set to deliver a much lower mean weight than machine B. For a machine delivering tens of thousands of packets per day the difference will mean a saving of tons of product per day if machine A is used. It is thus very important to have a measure of the spread (dispersion) of a distribution as well as its mean.

93

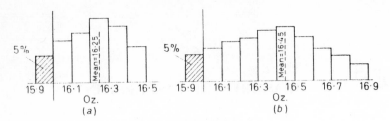

Figure 6.2

6.2. THE RANGE

The *range* is the simplest measure of dispersion. It is the difference between the largest and smallest values of the variate and takes no account of those in between. Thus the range of the set of numbers 24, 24, 27, 26, 22, 24, 25, 21, 22, 25 is $27 - 21 = 6$.

The range is in general a very poor measure of the dispersion, since it depends on the two extreme values of the variate, which are the least reliable. It has its biggest use in quality control (see Chapter 10) where it is assumed that the distribution is 'normal'.

6.3. THE MEAN DEVIATION

A second measure of dispersion would seem to be the mean of the deviations of a set of variates x_i ($i = 1, 2, \ldots n$) from their mean μ. That is $\sum_{i=1}^{N} (x_i - \mu)/N$ or $\sum_{i=1}^{n} f_i(x_i - \mu)/N$ allowing for various values of the variate to be repeated.

However, since $\sum_{i=1}^{n} f_i(x_i - \mu) = \sum_{i=1}^{n} f_i x_i - \sum_{i=1}^{n} f_i \mu$

$$= \sum_{i=1}^{n} f_i x_i - \mu \sum_{i=1}^{n} f_i$$

$$= \sum_{i=1}^{n} f_i x_i - \frac{\sum_{i=1}^{n} f_i x_i}{\sum_{i=1}^{n} f_i} \cdot \sum_{i=1}^{n} f_i$$

$$= \sum_{i=1}^{n} f_i x_i - \sum_{i=1}^{n} f_i x_i$$

$$= 0$$

the mean of the distances of the variates from the mean (all taken as positive) is used. This gives the mean deviation.

$$\text{Mean deviation} = \frac{\sum\limits_{i=1}^{n} f_i |x_i - \mu|}{\sum\limits_{i=1}^{n} f_i} \qquad \dots (6.1)$$

Note: If the set of variates is a sample and μ is unknown, then $\bar{x}$ replaces μ in the above formula.

Example 6.1. Find the mean deviation about the arithmetic mean of the following set of numbers 15, 17, 20, 24, 25, 26, 26, 30, 31.

Arithmetic mean

$$\mu = \frac{15 + 17 + 20 + 24 + 25 + 26 + 26 + 30 + 31}{9} = 23\cdot78.$$

The deviations about μ are: $(23\cdot78 - 15)$, $(23\cdot78 - 17)$, $(23\cdot78 - 20)$, $(24 - 23\cdot78)$, $(25 - 23\cdot78)$, ... $(31 - 23\cdot78)$.

Mean deviation

$$= \frac{8\cdot78 + 6\cdot78 + 3\cdot78 + 0\cdot22 + 1\cdot22 + 2\cdot22 + 2\cdot22 + 6\cdot22 + 7\cdot22}{9}$$

$$= \frac{38\cdot66}{9}$$

$$= 4\cdot3 \text{ (correct to 2 significant figures)}$$

Exercises 6a

1. A soap manufacturer uses machines which pack automatically 16 oz of soap flakes. To check on the average weight delivered sample packets are occasionally selected from the output and weighed accurately. The results of checking two machines are given in *Table 6.1*.

Table 6.1

Machine A	16·05	16·10	16·06	16·06	16·04	16·07	16·04	16·06	16·04	16·08
Machine B	16·03	16·10	16·13	16·08	16·08	16·05	15·99	16·12	16·03	15·99

Find the mean weight of each sample of 10 and compare the variability of the two machines by calculating the mean deviations from the means for each sample.

2. Calculate the mean deviation about the mean of the following set of numbers. 19, 10, 11, 21, 16, 13, 18, 17, 10. What would the mean deviation be if: (a) the numbers were each increased by 3, (b) the

numbers were each decreased by 2, (c) the numbers were each multiplied by 5, (d) the numbers were each divided by 2.

3. Calculate the median of the distribution of the values of 140 resistors given in *Table 6.2*. Hence find the mean deviation about the median. (Note the calculations for the mean deviations about the *median* are the same as in Example 6.1 except that the median is used in place of the mean.)

Table 6.2

$x \, \Omega$	f
45·0–46·0	3
–47·0	8
–48·0	14
–49·0	18
–50·0	27
–51·0	23
–52·0	22
–53·0	17
–54·0	6
–55·0	2
Total	140

4. In *Table 6.3* are the index numbers of production in the textile and food industries for the 10 years between 1924 and 1934, excluding 1926.

Table 6.3

Textiles	100	122	120	112	114	90	89	101	109	112
Food	100	97	103	103	108	109	113	112	110	110

Compare the variability in the two industries by calculating the mean deviation from the median for each set of index numbers.   (J.M.B.)

6.4. THE VARIANCE

Because of the modulus sign which was used in the mean deviation and the consequent awkward algebraic manipulations the mean deviation is not easy to use. A far more useful measure is the *variance* which uses the squared deviations $(x - \mu)^2$, which are all positive and hence do not cancel each other out. We define variance by

$$\text{Variance } (\sigma^2) = \frac{\sum\limits_{i=1}^{n} f_i(x_i - \mu)^2}{\sum\limits_{i=1}^{n} f_i} \qquad \ldots (6.2)$$

This has the dimensions of the square of the variate and hence is denoted by σ^2. Note the use of the Greek letter σ since the summation is taken throughout the whole population.

Use is also made of the *standard deviation* which is the square root of the variance

$$\text{Standard deviation } (\sigma) = \sqrt{\left(\frac{\sum\limits_{i=1}^{n} f_i(x_i - \mu)^2}{\sum\limits_{i=1}^{n} f_i}\right)} \qquad \ldots (6.3)$$

The variance can also be expressed in terms of expectation (see section 3.6)

$$\sigma^2 = \frac{\sum\limits_{i=1}^{n} f_i(x_i - \mu)^2}{\sum\limits_{i=1}^{n} f_i} = \frac{\sum\limits_{i=1}^{n} f_i(x_i - \mu)^2}{N}$$

$$= \sum\limits_{i=1}^{n} \frac{f_i}{N}(x_i - \mu)^2 = \sum\limits_{i=1}^{n} p_i(x_i - \mu)^2 = \mathscr{E}(x - \mu)^2 \qquad \ldots (6.4)$$

This is the most widely used measure of dispersion and because it takes into account all members of the population and can be manipulated algebraically (and hence arithmetically) it is one of the most satisfactory from the point of view of further statistical theory.

$$\text{Consider } \sigma^2 = \frac{\Sigma f(x - \mu)^2}{\Sigma f} \text{ (the subscript 'i' is understood)}$$

$$= \frac{\Sigma(f x^2 - 2fx\mu + f\mu^2)}{\Sigma f}$$

$$= \frac{\Sigma f x^2}{\Sigma f} - 2\mu \frac{\Sigma f x}{\Sigma f} + \mu^2 \frac{\Sigma f}{\Sigma f}$$

$$\text{but } \frac{\Sigma fx}{\Sigma f} = \mu \text{ the population mean}$$

$$\therefore \sigma^2 = \frac{\Sigma fx^2}{\Sigma f} - 2\mu\mu + \mu^2$$

$$= \frac{\Sigma fx^2}{\Sigma f} - \mu^2 \qquad \qquad \dots (6.5)$$

$$= \frac{\Sigma fx^2}{\Sigma f} - \left(\frac{\Sigma fx}{\Sigma f}\right)^2$$

$$\text{but } N = \Sigma f, \therefore \sigma^2 = \frac{1}{N}\left(\Sigma fx^2 - \frac{(\Sigma fx)^2}{N}\right) \qquad \dots (6.6)$$

This form is best used for arithmetical calculation.

Note: from equation (6.6)

$$\sigma^2 = \left(\sum \frac{f}{N}x^2\right) - \left(\sum \frac{f}{N}x\right)^2 = \mathscr{E}(x^2) - [\mathscr{E}(x)]^2 \quad \dots (6.7)$$

Example 6.2. Find the variance of the following numbers 310, 330, 370, 410, 420, 430, 450, 480.

One way of working is given in *Table 6.4*.

Table 6.4

x	f	fx	fx²
310	1	310	96,100
330	1	330	108,900
370	1	370	136,900
410	1	410	168,100
420	1	420	176,400
430	1	430	184,900
450	1	450	202,500
480	1	480	230,400
	$\Sigma f = 8$	$\Sigma fx = 3200$	$\Sigma fx^2 = 1,304,200$

From formula (6.6), $\sigma^2 = \dfrac{1}{8}\left(1{,}304{,}200 - \dfrac{(3200)^2}{8}\right) = \dfrac{24{,}200}{8} = 3025$

Note: standard deviation $\sigma = 55$.

The arithmetic used in working Example 6.2 is very lengthy and we recall that in Chapter 5 when calculating the arithmetical mean, use was made of a false origin 'a' and any common factor 'C' in order to lighten the working.

$$\text{If } d = \frac{x - A}{C}$$

$$x = A + Cd \qquad \qquad \dots \text{(i)}$$
$$\text{and } \mu = A + C\bar{d} \qquad \dots \text{(ii)}$$

where $\bar{d}$ is the mean of the d's $\left(\bar{d} = \dfrac{\Sigma fd}{\Sigma f}\right)$.

$$\therefore x - \mu = C(d - \bar{d}) \text{ from (i) and (ii)}.$$

Substituting this in equation (6.2) we have

$$\text{Variance } \sigma^2 = \frac{\Sigma f C^2 (d - \bar{d})^2}{\Sigma f}$$

$$= \frac{C^2}{N} \Sigma f(d - \bar{d})^2 \quad (C \text{ is constant})$$

$$= \frac{C^2}{N} \Sigma (fd^2 - 2fd\bar{d} + f\bar{d}^2)$$

$$= \frac{C^2}{N} (\Sigma fd^2 - 2\bar{d}\Sigma fd + \bar{d}^2\Sigma f)$$

$$= \frac{C^2}{N} (\Sigma fd^2 - 2\bar{d}N\bar{d} + N\bar{d}^2)$$

$$= \frac{C^2}{N} (\Sigma fd^2 - N\bar{d}^2)$$

$$= \frac{C^2}{N} \left[\Sigma fd^2 - \frac{(\Sigma fd)^2}{N}\right] \qquad \dots (6.8)$$

We shall now rework Example 6.2, using this new formula.

Example 6.3. Using the data of Example 6.2 but working with a false origin $A = 410$ and removing a common factor $C = 10$ so that $d = (x - 410)/10$ we have the results as given in *Table 6.5*.

Table 6.5

x	f	$d = \dfrac{x - 410}{10}$	fd	fd^2
310	1	-10	-10	100
330	1	-8	-8	64
370	1	-4	-4	16
410	1	0	$\underline{-22}$	0
420	1	$+1$	1	1
430	1	$+2$	2	4
450	1	$+4$	4	16
480	1	$+7$	7	49
$\Sigma f = 8$			$\Sigma fd = \dfrac{14}{-8}$	$\Sigma fd^2 = 250$

$$\text{From formula (6.8) } \sigma^2 = \frac{100}{8}\left[250 - \frac{(-8)^2}{8}\right]$$

$$= \frac{100}{8} \times 242$$

$$= 3025 \text{ (as before)}$$

The next example shows the same sort of calculation applied to a grouped distribution.

Example 6.4

In *Table 6.6*, columns (*vii*), (*viii*) and (*ix*) are worked to check the result. Two checks are available.

(*a*) Since $\Sigma f(d + 1) - \Sigma fd = \Sigma f$ (see also Example 5.3).

sum of column (viii) − sum of column (v) = sum of column (ii).

In this case $304 - 166 = 138$

(*b*) Since $\Sigma f(d + 1)^2 - \Sigma fd^2 - 2\Sigma fd = \Sigma f$
sum of column (*ix*) − sum of column (*vi*) − 2 × sum of column (*v*)
= sum of column (*ii*).
In this case $3022 - 2552 - 2 \times 166 = 138$

Table 6.6

x	f	Mid-interval values	$d = \dfrac{x-28}{4}$	fd	fd^2	$(d+1)$	$f(d+1)$	$f(d+1)^2$
(i)	(ii)	(iii)	(iv)	(v)	(vi)	(vii)	(viii)	(ix)
0–8	3	4	−6	−18	108	−5	−15	75
−16	17	12	−4	−68	272	−3	−51	153
−24	24	20	−2	−48	96	−1	−24	24
−32	29	28	0	−134	0	1	29	29
−40	27	36	2	54	108	3	81	243
−48	16	44	4	64	256	5	80	400
−56	11	52	6	66	396	7	77	539
−64	5	60	8	40	320	9	45	405
−80	4	72	11	44	484	12	48	576
−104	2	92	16	32	512	17	34	578

$\Sigma f = 138$

$+300$, $\Sigma fd = 166$

$\Sigma fd^2 = 2552$

$\Sigma f(d+1) = 394 - 90 = 304$

$\Sigma f(d+1)^2 = 3022$

Since $\Sigma f = 138$, $\Sigma fd = 166$, $\Sigma fd^2 = 2552$, $C = 4$ from formula (6.7)

$$\text{Variance} = \frac{4^2}{138}\left(2552 - \frac{166^2}{138}\right)$$

$$= 272 \cdot 7.$$

In practice we generally have only a sample of the population and have to estimate the population variance from the sample. In section 3.8 we saw that $\bar{x}$, the sample mean, is an unbiased estimate of μ, the population mean, i.e. $\mathscr{E}(\bar{x}) = \mu$.

It is not true, however, that the sample variance $[\sum_{i=1}^{n} f_i(x_i - \bar{x})^2]/N$ is an unbiased estimate of the population variance σ^2. In Chapter 14 it is shown that:

$$\mathscr{E}\left[\frac{\sum_{i=1}^{n} f_i(x_i - \bar{x})^2}{N - 1}\right] = \sigma^2$$

Thus on average $s^2 = [\Sigma f(x - \bar{x})^2]/(N - 1)$ gives a better estimate of the population variance. Henceforth in this book the sample variance will be taken as given by s^2 as defined above.

Note: however: If we are dealing with a sample and μ is known and replaces $\bar{x}$ then $s^2 = [\Sigma f(x - \mu)^2]/N$ (i.e. 'N' replaces '$N - 1$').

Example 6.5. If the data given in Example 6.2 are a random sample from a population, estimate the variance σ^2 of the parent population.

As we now have a sample and μ is unknown we estimate σ^2 by using

$$s^2 = \frac{\Sigma f(x - \bar{x})^2}{N - 1} = \frac{24{,}200}{7} = 3457 \cdot 1$$

Example 6.6. Find the variance of the Poisson distribution (see sections 4.3, 4.4).

Consider:

$$e^{a(t-1)} = e^{-a}e^{at} = e^{-a}\left(1 + at + \frac{a^2t^2}{2!} + \frac{a^3t^3}{3!} + \dots\right)$$

Differentiate both sides with respect to t:

$$ae^{a(t-1)} = e^{-a}\left(0 + a + 2.\frac{a^2t}{2!} + 3.\frac{a^3t^2}{3!} + \dots\right) \qquad \dots\text{(i)}$$

If we let $t = 1$ in this equation we have that the mean of the Poisson distribution is a, also $\mathscr{E}(x) = a$ (see section 4.4).

Multiply equation (i) by t

$$ate^{a(t-1)} = e^{-a}(at + \frac{2a^2t^2}{2!} + \frac{3a^3t^3}{3!} + \ldots)$$

Differentiating this equation with respect to t we have

$$ae^{a(t-1)} + a^2te^{a(t-1)} = e^{-a}(a + \frac{2^2a^2t}{2!} + \frac{3^2a^3t^2}{3!} + \ldots)$$

Now let $t = 1$ (remembering that $e^0 = 1$).

$$a + a^2 = \frac{a}{1!}e^{-a} + 2^2\frac{a^2}{2!}e^{-a} + 3^2\frac{a^3}{3!}e^{-a} + \ldots$$

$$= 0\,P_0 + 1^2\,P_1 + 2^2\,P_2 + 3^3\,P_3 + \ldots$$

where $P_0, P_1, P_2, \ldots$ are the probabilities of the Poisson distribution.

$$\text{Thus } a + a^2 = \sum_0^\infty x^2 P_x$$

$$a = \sum_0^\infty x^2 P_x - a^2$$

$$= \mathscr{E}(x^2) - (\mathscr{E}(x))^2$$

$$= \sigma^2 \text{ (see equation (6.7))}$$

Note that for a Poisson distribution $\mu = a = \sigma^2$.

Example 6.7. Find the variance of the binomial distribution (see sections 4.1 and 4.2.). Consider

$$(q + pt)^n = q^n + {}^nC_1q^{n-1}pt + {}^nC_2q^{n-2}p^2t^2 + \ldots + p^nt^n$$

$$= P_0 + P_1t + P_2t^2 + \ldots P_nt^n \qquad \ldots (i)$$

where $P_0, P_1, P_2, \ldots P_n$ are the probabilities of the binomial distribution. Differentiate both sides of equation (i) with respect to t:

$$np(q + pt)^{n-1} = 0 + P_1 + 2P_2t + 3P_3t^2 + \ldots + nP_nt^{n-1}.$$

If we let $t = 1$ in this equation we have that the mean of the binomial distribution is np and also

$$\mathscr{E}(x) = np \qquad \ldots (ii)$$

(see section 4.2)
Multiply equation (i) by t.

$$npt(q + pt)^{n-1} = P_1t + 2P_2t^2 + 3P_3t^3 + \ldots + nP_nt^n$$

E

Differentiating this equation with respect to t we have

$$np(q + pt)^{n-1} + n(n - 1)p^2t(q + pt)^{n-2} = P_1 + 2^2P_2t + 3^2P_3t^2 + \ldots$$
$$+ n^2P_nt^{n-1}$$

Now let $t = 1$ (remembering that $q + p = 1$), then

$$np + n(n - 1)p^2 = 0P_0 + 1^2P_1 + 2^2P_2 + 3^2P_3 + \ldots + n^2P_n$$

$$np(1 + np - p) = \sum_0^n x^2P_x$$

$$np(1 - p) + n^2p^2 = \mathscr{E}(x^2)$$

$$npq \qquad\qquad = \mathscr{E}(x^2) - (np)^2$$

$$\qquad\qquad = \mathscr{E}(x^2) - [\mathscr{E}(x)]^2 \text{ see equation (ii)}$$

$$\qquad\qquad = \sigma^2 [\text{see equation (6.7)}]$$

that is the variance of the binomial distribution $= npq \ldots . (6.9)$

Exercises 6b

1. Using the two samples given in question 1, Exercises 6a estimate the standard deviation of the total output of the two packing machines.

2. Find the variance of the n natural numbers $1, 2, 3, \ldots n$.

3. The standard deviation of a set of numbers is to be σ. On checking the data a systematic error is found. Without referring to the original data, what is the true value of the standard deviation in the following cases: (a) Each number is 2·0 too high. (b) Each number is 2·0 too low. (c) Half the numbers are 2·0 too high and the other half 2·0 too low. (d) Each number is twice as large as it should be.

4. The noonday temperature at Quito (Ecuador) was noted for every day of one year. The arithmetic mean of the distribution of noonday temperatures was found to be 60°F. Why is it not true to say that all places with an average yearly noonday temperature of 60°F have a similar climate to that of Quito?

5. Find the variance of the distributions of heights and weights of 100 students given in Table 6.7.

6.5. THE COEFFICIENT OF VARIATION

Consider the following:

(a) A standard deviation of 1 ft in the measurements of the lengths of planks whose average length is 100 ft,

(b) the same variation of 1 ft in the measurements of planks whose average length is 5 ft.

Obviously the spread about the mean of the lengths in case (a) is less important than the spread in case (b). The *coefficient of variation* can be used to give some measure of the relative importance of the standard

Table 6.7

(a)			(b)	
Height (in) (measured to the nearest 0·1 in)	f		Weight (lb) (measured to the nearest lb)	f
60·0–61·4	2		110–119	1
61·5–62·9	7		120–129	3
63·0–64·4	10		130–139	7
64·5–65·9	17		140–149	16
66·0–67·4	19		150–159	21
67·5–68·9	15		160–169	18
69·0–70·4	13		170–179	12
70·5–71·9	9		180–189	10
72·0–73·4	5		190–199	8
73·5–74·9	3		200–210	4
Total	100		Total	100

deviation referred to the mean. We define the

$$\text{Coefficient of variation} = \frac{\text{Standard deviation}}{\text{Mean}} \quad \dots (6.10)$$

Note: the coefficient of variation is not of great value when the mean is very small in absolute size.

Exercises 6c

1. Calculate the coefficients of variation for the heights and weights of the students in Exercises 6b question 5.

2. In an industry A the average monthly wage is 1210 s. with a standard deviation of 24 s. In a second industry B the average monthly wage is 1090 s. with a standard deviation of 24 s. Compare the variation of wages in the two industries by calculating the coefficients of variation.

EXERCISES 6

1. Find the standard deviation of the distribution of the life of 100 electric light bulbs given in Table 6.8.

Table 6.8

Life (h)	0–250	–500	–750	–1000	–1250	–1500	–1750	–2000
Frequency	1	3	7	12	25	39	11	2

2. A set of N_1 variates has a mean μ_1 and variance σ_1^2. Another set of N_2 variates has a mean μ_2 and variance σ_2^2. The two sets of numbers are combined into a single set; if the combined set has a mean μ prove that its variance is given by

$$\sigma^2 = \frac{N_1\sigma_1^2 + N_2\sigma_2^2 + N_1(\mu_1 - \mu)^2 + N_2(\mu_2 - \mu)^2}{N_1 + N_2}$$

3. Two samples each of 25 items are taken from the output of an automatic machine. The mean and standard deviation of the first sample are found to be 2·000 and 0·030 respectively, and of the second sample 2·002 and 0·033 respectively. Find the mean and standard deviation of the combined sample of 50 items.

4. Find the mode, median, mean, range, mean deviation about the mean and the standard deviation for the two distributions given in Table 6.9.

Table 6.9

(a)	x	0	1	2	3	4	5	6	7	8	9
	f	2	6	12	9	19	39	42	39	21	11

(b)	x	3·00–3·10	–3·20	–3·30	–3·40	–3·50	–3·60	–3·70	–3·80
	f	16	25	23	12	10	7	5	2

5. Compare the skewness of the two distributions given in the previous question by means of Pearson's Measure of Skewness, $\dfrac{\text{mean} - \text{mode}}{\text{standard deviation}}$. Also find the coefficient of variation for each distribution.

6. A set of numbers are all either 0 or 1. If m of the numbers are 1 and n of the numbers are 0, prove that the variance of the set of numbers is $\dfrac{m \times n}{(m + n)^2}$.

7. A set of N numbers $x_1, x_2, \ldots x_n$ are all different. If $\Sigma x^2 = 3600$, $\Sigma x = 200$ and $\sigma^2 = 80$, where σ^2 is the variance of the set, find N.

8. Two distributions have population sizes N_1 and N_2, means μ_1 and μ_2 and standard deviations σ_1^2 and σ_2^2 respectively. Show that the variance of the combined distribution is σ^2 where

$$(N_1 + N_2)\sigma^2 = N_1\sigma_1^2 + N_2\sigma_2^2 + \frac{N_1 N_2}{N_1 + N_2}(\mu_1 - \mu_2)^2.$$

9. From *Table 6.10* find the mean and standard deviation distributions. Plot a cumulative frequency graph and from it find the 75 percentile.

Table 6.10

x	f
4·30–4·40	4
–4·50	8
–4·60	8
–4·70	9
–4·80	6
–4·90	0
–5·00	3
–5·10	1
Total	39

10. Given that $\sum_{r=1}^{n-1} r = \frac{1}{2} n(n - 1)$, and $\sum_{r=1}^{n-1} r^2 = \frac{n}{6}(n - 1)(2n - 1)$, prove that the variance of the terms of an arithmetic progression of n terms $a, a + d, a + 2d, \ldots a + (n - 1)d$ is $\frac{1}{12}(n^2 - 1)d^2$.

11. If $N = \Sigma f$, $\bar{x} = (\Sigma fx)/N$, $\mu_2(a) = (1/N) \Sigma f(x - a)^2$, $\mu_3(a) = (1/N) \Sigma f(x - a)^3$; $\mu_4(a) = (1/N) \Sigma f(x - a)^4$, prove that: (a) $\mu_3(\bar{x}) = \mu_3(a) - 3(\bar{x} - a)\mu_2(a) - 2(\bar{x} - a)^3$, (b) $\mu_4(\bar{x}) = \mu_4(a) - 4(\bar{x} - a)\mu_3(a) - 6(\bar{x} - a)^2\mu_2(a) - 3(\bar{x} - a)^4$.

Use these relations to evaluate $\mu_3(\bar{x})$ and $\mu_4(\bar{x})$ for the distribution of heights of compression springs (take a as $1·065$ in) (*Table 6.11*).

Table 6.11

x	f
1·000–1·010	2
–1·020	5
–1·030	7
–1·040	9
–1·050	11
–1·060	14
–1·070	18
–1·080	13
–1·090	12
–1·100	9
Total	100

12. Assuming that the variance of the mean of a sample of size n drawn from a population with variance σ^2 is σ^2/n, show that the standard error of the percentage defective, in a sample from a binomial population of effective and defective components is, at most, $50/\sqrt{n}$.

A sample of 400 components from one batch, and 600 components from another are examined. Both samples represent a small proportion of batch size and the difference between their percentage defective is 6·5. Show that this is evidence suggesting that the batches have different percentages defective, whatever these may be.   (Liv.U.)

13. From the frequency distribution of a variable y are calculated: (a) mean, (b) median, (c) upper quartile, (d) variance, (e) standard deviation, (f) mean deviation from the mean, (g) semi-interquartile range. What is the relation between these quantities and the corresponding quantities for the frequency distribution of $ay + b$, where a and b are constants? Give proofs of all your statements.   (Liv. U.)

14. The lengths of 524 microfilaria in pleural blood were each measured to the nearest micron. See *Table 6.12*.

Calculate the mean and standard deviation of this distribution.

                                                (Liv. U.)

15. *Table 6.13* gives the frequency distribution of the times, to the

Table 6.12

Length (μ)	Frequency	Length (μ)	Frequency
35–39	1	60–64	39
40–44	2	65–69	123
45–49	3	70–74	172
50–54	10	75–79	117
55–59	13	80–84	40
		85–89	4

Table 6.13

Time min	sec	min	Frequency sec	
7	10	– 7	19	4
7	20	– 7	29	8
7	30	– 7	39	8
7	40	– 7	49	9
7	50	– 7	59	6
8	0	– 8	9	0
8	10	– 8	19	3
8	20	– 8	29	1

nearest second, of the 39 races (heats and finals) for the 'Thames Challenge Cup' at the Henley Regatta, 1956.

Plot the cumulative frequency graph and use it to determine

(a) the 80th percentile,

(b) the percentile rank of a time of 7 min 45 sec.

Explain the meaning of your results. Calculate the mean time and the standard deviation. (J.M.B.)

16. A sample of 100 deaths registered during a particular period

(see *Table 6.14*) had the following distribution of age at the last birthday, in years.

Table 6.14

Age at death	0–9	10–19	20–29	30–39	40–49	50–59	60–69	70–79	80–89
Frequency	14	1	2	6	5	15	17	21	19

Find the mean m and the standard deviation s, and evaluate the coefficient of variation α given by $\alpha = 100 \, s/m$.

Give a brief explanation of the double-peaked nature of the distribution. (J.M.B.)

7

CONTINUOUS DISTRIBUTIONS

7.1. INTRODUCTION

IT was seen in Chapter 2 how we could group into classes, values which were measured on a continuous scale, and present them as a histogram or frequency polygon. If we alter the vertical scale of such a histogram

$$\text{from} \quad \frac{\text{frequency}}{\text{class width}} \quad \text{to} \quad \frac{\text{frequency}}{\text{class width} \times \text{total frequency}} \quad \dots \text{(i)}$$

a histogram is obtained in which the total area is equal to unity.

$$\text{The area above a given class width} = \frac{\text{frequency}}{\text{total frequency}} \begin{pmatrix} \text{multiply (i) by} \\ \text{class width} \end{pmatrix}$$

$$= \text{probability}$$

That is, the area above a given class width gives the probability that a value of the variate will come in that class. In this form we have a *relative frequency histogram*. For example, referring to *Figure 2.5* we can make this a relative frequency histogram by dividing the vertical scale by the total frequency (in this case 100).

If we let our sample increase in size we can make our intervals smaller and the outline of the relative frequency histogram will approximate to a curve $y = h(x)$, Oy being the vertical scale and Ox the horizontal scale. If x extends from a to b we have that the area under the curve is $\int_a^b h(x)\,dx = 1$.

With a continuous curve we consider a small element of area $h(x)\,dx$ to be the probability element associated with the point x. Thus $h(x)$ is called the *probability density function*.

The probability of x lying between x_1 and x_2 is given by $\int_{x_1}^{x_2} h(x)\,dx$.

Example 7.1. A variate x can assume only values between 0 and 8 and its probability density function is $kx(8 - x)$ $(0 \leqslant x \leqslant 8)$. Find the value of k and show that the probability of a random variate x being in the range $x = 0$ to $x = 2$ is $\frac{5}{32}$. Sketch the curve.

We recall that the total area under such a curve is unity

$$\therefore \int_0^8 kx(8 - x)\,dx \quad = 1$$

$$k\int_0^8 (8x - x^2)\,dx \quad = 1$$

$$k\left(4x^2 - \frac{x^3}{3}\right)_0^8 = 1$$

$$k\left(256 - \frac{512}{3}\right) - (0) = 1$$

$$k\left(\frac{256}{3}\right) = 1$$

$$k = \frac{3}{256}.$$

The probability of x in the range $x = 0$ to $x = 2$ is given by

$$\int_0^2 \frac{3}{256} x(8 - x)\,\mathrm{d}x = \frac{3}{256}\int_0^2 (8x - x^2)\,\mathrm{d}x$$

$$= \frac{3}{256}\left(4x^2 - \frac{x^3}{3}\right)_0^2$$

$$= \frac{3}{256}\left(16 - \frac{8}{3}\right) - (0)$$

$$= \frac{3}{256} \times \frac{40}{3}$$

$$= \frac{5}{32} \text{ (as required)}$$

This is usually written $\Pr(0 \leqslant x \leqslant 2) = \dfrac{5}{32}$.

In order to sketch the curve we find the number and nature of the turning points.

$$y = \frac{3}{256} x(8 - x)$$

$$\frac{256\,y}{3} = 8x - x^2$$

differentiating $$\frac{256}{3}\frac{\mathrm{d}y}{\mathrm{d}x} = 8 - 2x$$

$$\frac{256}{3}\frac{\mathrm{d}^2 y}{\mathrm{d}x^2} = -2$$

for turning points $$\frac{\mathrm{d}y}{\mathrm{d}x} = 0$$

i.e. $8 - 2x = 0$

thus $x = 4$ is the only turning point.

Substituting $x = 4$ in the original equation we find that $y = \frac{3}{16}$ and since d^2y/dx^2 is negative the point $(4, \frac{3}{16})$ is a maximum, also when $x = 0$ or 8, $y = 0$. The curve is as shown in *Figure 7.1*.

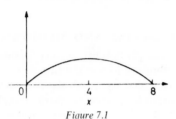

Figure 7.1

Example 7.2. Given that a variate x is distributed according to the probability density function $h(x) = x e^{-x}$ $(x \geqslant 0)$, find the probability of obtaining a value of x between 0 and 1. (*Note*: $\lim\limits_{x \to \infty} x^n e^{-x} = 0$.)

We first note that the limits of x are 0 and ∞, and verify that

$$\int_0^\infty h(x)\,dx = \int_0^\infty x e^{-x}\,dx = \lim_{R \to \infty} \left[\int_0^R x e^{-x} dx \right]$$

$$= \lim_{R \to \infty} \left[-x e^{-x} - e^{-x} \right]_0^R$$

$$= \lim_{R \to \infty} \left[-R e^{-R} - e^{-R} - (0 - 1) \right]$$

$$= 1$$

The required probability is given by

$$\int_0^1 x e^{-x}\,dx = \left[-x e^{-x} - e^{-x} \right]_0^1$$

$$= (-e^{-1} - e^{-1}) - (-0 - 1)$$

$$= 1 - \frac{2}{e}$$

$$= 1 - 0{\cdot}736$$

$$= 0{\cdot}264.$$

Exercises 7a

1. A variate x can assume only values between 0 and 5 and the equation of its relative frequency curve is $y = kx(5 - x)$. Find the value of k. Show that the probability of a value of x in the range $(0, \frac{5}{4})$ is $\frac{5}{32}$, and sketch the relative frequency curve.

2. A variate x can assume any positive value and the equation of

its relative frequency curve is $y = kx^2e^{-x}$. Show that $k = \frac{1}{2}$ and sketch the curve. (*Note*: $\lim\limits_{x \to \infty} x^n e^{-x} = 0$.)

3. A variate x can assume any positive value and the equation of its relative frequency curve is $y = ke^{-x}$. Show that $k = 1$ and the probability that a value of x is in the range $0 \cdot 288 \leqslant x \leqslant 1 \cdot 39$ is approximately $0 \cdot 5$. Sketch the curve.

7.2. THE MODAL AND MEDIAN VALUES

The modal value for the probability density function $h(x)$ is the value of the variate corresponding to a maximum value. It can be found by differentiation as illustrated in Example 7.3. The median value M divides the total area under the curve into two equal parts. Since the total area under the curve is unity the equation to be solved is either (assuming x is defined for $a \leqslant x \leqslant b$)

$$\int_a^x h(x)\,dx = \frac{1}{2} \text{ or } \int_x^b h(x)\,dx = \frac{1}{2} \qquad \dots (7.1)$$

An exact solution of such equations is not always possible, each one has to be treated on its own merits. See Examples 7.4 and 7.5.

Example 7.3. A relative frequency distribution is defined by the curve $y = 4(x - x^3)\,(0 \leqslant x \leqslant 1)$. Find its modal value.

$$y = 4x - 4x^3$$

$$\therefore \qquad \frac{dy}{dx} = 4 - 12x^2$$

$$\frac{d^2y}{dx^2} = -24x$$

Turning values are given by $\dfrac{dy}{dx} = 0$

i.e.
$$4 - 12x^2 = 0$$
$$12x^2 = 4$$
$$x = \pm \frac{1}{\sqrt{3}}$$

Since x lies between 0 and 1 we only consider $x = +\dfrac{1}{\sqrt{3}}$; for this value $\dfrac{d^2y}{dx^2}$ is negative, hence this value of x gives a maximum, and the required modal value is $\dfrac{1}{\sqrt{3}}$.

Example 7.4. Find the median X for the probability density function $h(x) = e^{-x} (x \geqslant 0)$.

$$\int_0^X e^{-x} dx = \tfrac{1}{2}$$

$\therefore$
$$[-e^{-x}]_0^X = \tfrac{1}{2}$$

$$1 - e^{-X} = \tfrac{1}{2}$$

$$e^{-X} = \tfrac{1}{2}$$

$$e^X = 2$$

$$X = \log_e 2 \text{ (note this gives an exact solution).}$$

Example 7.5. A population has a probability density function $xe^{-x} (x \geqslant 0)$. Find the median X.

$$\int_0^X xe^{-x} dx = \tfrac{1}{2}$$

$\therefore$
$$[-xe^{-x} + \int e^{-x} dx]_0^X = \tfrac{1}{2}$$
$$[-xe^{-x} - e^{-x}]_0^X = \tfrac{1}{2}$$
$$-Xe^{-X} - e^{-X} + 1 = \tfrac{1}{2}$$

i.e.
$$\tfrac{1}{2} = e^{-X}(1 + X).$$

This equation cannot be solved exactly; an approximate solution is $X = \tfrac{5}{3}$.

7.3. MATHEMATICAL EXPECTATION, THE MEAN AND THE VARIANCE

For a continuous probability density function $h(x)$ $(a \leqslant x \leqslant b)$ we define the expectation of x to be

$$\mathscr{E}(x) = \int_a^b xh(x) \, dx \qquad \dots (7.2)$$

Similarly
$$\mathscr{E}(x^n) = \int_a^b x^n h(x) \, dx \qquad \dots (7.3)$$

(refer also to section 3.6). The following results apply:

(a) $\mathscr{E}(k) = k$ (k constant)
(b) $\mathscr{E}(kx) = k\mathscr{E}(x)$ (k constant)
(c) $\mathscr{E}(x + y) = \mathscr{E}(x) + \mathscr{E}(y)$

The arithmetic mean $\mu = \mathscr{E}(x) = \int_a^b xh(x) \, dx$ $\qquad \dots (7.4)$

The variance of x $\text{var}(x) = \mathscr{E}(x - \mu)^2 = \int_a^b (x - \mu)^2 h(x)$ $\qquad \dots (7.5)$

From Chapter 14 $\mathscr{E}(x - \mu)^2 = \mathscr{E}(x^2) - (\mathscr{E}(x))^2$ $\qquad \dots (7.6)$

$\therefore$ $\text{Var}(x) = \int_a^b x^2 h(x) \, dx - \mu^2$ $\qquad \dots (7.7)$

Example 7.6. Find the mean of the distribution defined by the probability density function $4(x - x^3)$ $(0 \leqslant x \leqslant 1)$.

By the formula (7.4) $\mu = \int_0^1 x \, 4(x - x^3) \, dx$

$$= \int_0^1 (4x^2 - 4x^4) \, dx$$

$$= \left(\frac{4x^3}{3} - \frac{4x^5}{5} \right)_0^1$$

$$= \left(\frac{4}{3} - \frac{4}{5} \right) - (0)$$

$$= \frac{8}{15}$$

Example 7.7. Find the variance of a continuous variate x whose probability density function is given by xe^{-x} $(x \geqslant 0)$. Using the formula (7.7)

$$\text{var}(x) = \int_0^\infty x^2(xe^{-x}) \, dx - \mu^2 \qquad \ldots (i)$$

Now $\mu = \int_0^\infty x(xe^{-x}) \, dx$

$$= \lim_{R \to \infty} \int_0^R x^2 e^{-x} \, dx$$

$$= \lim_{R \to \infty} \left[-x^2 e^{-x} + \int 2xe^{-x} \, dx \right]_0^R \text{ (by parts)}$$

$$= \lim_{R \to \infty} \left[-R^2 e^{-R} + 0 + 2[-xe^{-x} + \int e^{-x} \, dx]_0^R \right]$$

$$= \lim_{R \to \infty} \left[-R^2 e^{-R} + 0 + 2(-Re^{-R} + 0 - e^{-R} + 1) \right]$$

$$= 0 + 2(0 + 1)$$

$$= 2 \qquad \ldots (ii)$$

Returning to equation (i)

$$\int_0^\infty x^2(xe^{-x}) \, dx = \int_0^\infty x^3 e^{-x} \, dx$$

$$= \lim_{R \to \infty} \{ [-x^3 e^{-x}]_0^R + \int_0^R 3x^2 e^{-x} \, dx \}$$

$$= 0 + 3 \times 2 \text{ (refer to the above calculation for } \mu)$$

$$= 6 \qquad \ldots (iii)$$

From (i), (ii) and (iii) $\text{var}(x) = 6 - 2^2 = 2$.

7.4. THE MEAN DEVIATION ABOUT THE MEAN

If μ denotes the mean then the mean deviation about the mean for the probability density function $h(x)$ $(a \leqslant x \leqslant b)$ is given by

$$\int_b^a |x - \mu| \, h(x) \, dx \qquad \qquad \dots (7.8)$$

μ lies between a and b and the range of integration can be split into two parts

      (a) $a \rightarrow \mu$ where $x < \mu$ therefore $|x - \mu| = \mu - x$
      (b) $\mu \rightarrow b$ where $x > \mu$ therefore $|x - \mu| = x - \mu$

the expression (7.8) becomes

$$\int_a^\mu (\mu - x) \, h(x) \, dx + \int_\mu^b (x - \mu) \, h(x) \, dx \qquad \dots (7.9)$$

Example 7.8. Find the mean deviation from the mean for the continuous variate x whose probability density function is given by $e^{-x}(x \geqslant 0)$.

$$\begin{aligned}
\text{Mean} &= \int_0^\infty x \, e^{-x} \, dx \; [\text{see formula (7.4)}] \\
&= [- x e^{-x} - e^{-x}]_0^\infty \\
&= (-0 - 0) - (-0 - 1) \\
&= 1
\end{aligned}$$

Using formulae (7.8) and (7.9)

$$\begin{aligned}
\text{Mean deviation} &= \int_0^\infty |x - 1| \, e^{-x} \, dx \\
&= \int_0^1 (1 - x) e^{-x} \, dx + \lim_{R \to \infty} \left[\int_1^R (x - 1) e^{-x} \, dx \right] \\
&= [-(1 - x) e^{-x} + e^{-x}]_0^1 + \lim_{R \to \infty} [-(x - 1) e^{-x} - e^{-x}]_1^R \\
&= (-0 + e^{-1}) - (-1 + 1) + (-0 - 0) - (0 - e^{-1}) \\
&= 2 e^{-1} \\
&= 0.74 \text{ (approx.)}
\end{aligned}$$

Exercises 7b

1. A distribution has the probability density function e^{-x} $(x \geqslant 0)$. Verify that the median value is approximately 0.693 and find the variance.

2. Find the modal values of the distributions with the following probability density functions
    (a) $x e^{-x}(x \geqslant 0)$,
    (b) $4/27 \, (-9 + 12x - x^3) \, (0 \leqslant x \leqslant 3)$.

3. Find the means of the distributions whose probability density

functions are:

  (a) xe^{-x} $(x \geqslant 0)$

  (b) $\frac{1}{9}(3 + 2x - x^2)$ $(0 \leqslant x \leqslant 3)$

  (c) $\frac{3}{32}(4x - x^2)$ $(0 \leqslant x \leqslant 4)$

In the last case (c) was equation (7.4) necessary in order to find the mean?

4. Find the mean deviation from the mean for the continuous variate x whose probability density function is:

  (a) $xe^{-x} (x \geqslant 0)$ [refer to the answer for question 3(a)]

  (b) $1/a \, (0 \leqslant x \leqslant a) \, (a$ constant)

5. If x has the probability density function $kx(2 - x) \, (0 \leqslant x \leqslant 2)$ find k and the mean and variance of the distribution.

6. Find the value of Pearson's measure of skewness,

$$\frac{\text{mean} - \text{mode}}{\text{standard deviation}}$$

for the variate x whose probability density function is $xe^{-x} \, (x \geqslant 0)$ (use the results of question 2(a) and Example 7.7).

7.5. THE RECTANGULAR DISTRIBUTION

The *rectangular distribution* is a simple example of a continuous distribution defined by a probability density function.

The rectangular distribution is

$$h(x) = 1/A \quad (0 \leqslant x \leqslant A) \qquad \qquad \ldots (7.10)$$

A is a constant. The range of values of x has been taken from 0 to A, in fact they could have been taken over any range of width A. The distribution is shown in *Figure 7.2*.

The distribution has no mode and by symmetry its mean and median are both $A/2$.

Its variance, $\text{var}(x) = \mathscr{E}(x^2) - (\mathscr{E}(x))^2$ (see Chapter 14)

$$= \int_0^A x^2 \frac{1}{A} \, dx - (\mathscr{E}(x))^2$$

$$= \left[\frac{x^3}{3A} \right]_0^A - (\text{mean})^2$$

$$= \frac{A^2}{3} - \frac{A^2}{4}$$

$$= \frac{A^2}{12} \qquad \qquad \ldots (7.11)$$

Example 7.9. The length x of a side of a square is rectangularly distri-

Figure 7.2. The rectangular distribution

buted between 3 and 5. Show that the area of the square is distributed between 9 and 25 with a probability density function $\frac{1}{4} y^{-\frac{1}{2}}$. Calculate the mean and variance of the area of the square.

Since the range of x is 2 units, $h(x) = \frac{1}{2} (3 \leqslant x \leqslant 5)$.

$$\text{Now } 1 = \int_3^5 h(x)\,dx$$

$$= \int_3^5 \tfrac{1}{2}\,dx$$

but $x = y^{\frac{1}{2}}$ therefore $dx = \frac{1}{2} y^{-\frac{1}{2}}\,dy$, and when $x = 3$, $y = 9$ and when $x = 5$, $y = 25$.

$$\therefore \quad 1 = \int_9^{25} \tfrac{1}{2} \cdot \tfrac{1}{2} y^{-\frac{1}{2}}\,dy$$

$$= \int_9^{25} \tfrac{1}{4} y^{-\frac{1}{2}}\,dy.$$

Thus the area is distributed with a probability density function $\frac{1}{4} y^{-\frac{1}{2}}$ $(9 \leqslant y \leqslant 25)$.

To find the mean

$$\mathscr{E}(y) = \int_9^{25} y \times \tfrac{1}{4} y^{-\frac{1}{2}}\,dy$$

$$= \int_9^{25} \tfrac{1}{4} y^{\frac{1}{2}}\,dy$$

$$= \left[\tfrac{1}{6} y^{\frac{3}{2}} \right]_9^{25}$$

$$= \tfrac{1}{6}(125 - 27)$$

$$= 16\tfrac{1}{3}$$

To find the variance

$$\text{var}(x) = \mathscr{E}(y^2) - [\mathscr{E}(y)]^2$$

$$= \int_9^{25} y^2 \times \tfrac{1}{4} y^{-\frac{1}{2}}\,dy - [\mathscr{E}(y)]^2$$

$$= \int_9^{25} \tfrac{1}{4} y^{\frac{3}{2}}\,dy - (16\tfrac{1}{3})^2$$

$$= \left[\frac{1}{10} y^{\frac{5}{2}} \right]_9^{25} - (16\tfrac{1}{3})^2$$

$$= \frac{1}{10}(3125 - 243) - \frac{2401}{9}$$

$$= \frac{2882}{10} - \frac{2401}{9}$$

$$= 21 \cdot 4$$

EXERCISES 7

1. A variate x can assume any value between 0 and 4 and the equation of its relative frequency curve is $y = k(6 - x)(x + 2)$. Find the value of k and show that the relative frequency in the range $x = 0$ to $x = 2$ is 0·5.

2. A variate x can assume any value between 4 and 10 and the equation of its relative frequency curve is $y = k$ (k a constant). State the value of k and sketch the frequency curve.

3. A variate x can assume any value between -1 and $+1$, and the equation of its relative frequency curve is $y = k/(1 + x^2)$. Show that $k = 2/\pi$ and sketch the relative frequency curve.

4. A variate x has the probability density function $3x^2$ $(0 \leqslant x \leqslant 1)$. Find $\mathscr{E}(x)$ and $\mathscr{E}(x^2)$.

5. A variate x has the probability density function $x/2$ $(0 \leqslant x \leqslant 2)$. If two values of x between 0 and 2 are picked at random what is the probability that both are greater than one?

6. The length x of a cube is rectangularly distributed between 0 and 3. Show that the volume V of the cube is distributed between 0 and 27 with the probability density function $\frac{1}{6} V^{-\frac{2}{3}}$. Make a sketch of the distribution and calculate its mean and variance.

7. The radius of a circle is rectangularly distributed between 1 and 2. Show that the area A of the circle is distributed between π and 4π with a probability density function of $\frac{1}{2} \pi^{-\frac{1}{2}} A^{-\frac{1}{2}}$. Sketch the relative frequency curve and calculate the mean and variance of the area of the circle.

8. Find the mode and the mean of the distribution, the equation of whose probability density function is

$$\tfrac{2}{27}(6 + x - x^2) \quad (0 \leqslant x \leqslant 3)$$

Also show that the median is approximately 1·1.

9. A variate x can assume values only between 0 and 1 and the equation of its probability density function is $k\,e^{-3x}$. Find the value of k to three decimal places, and also the mean of the distribution.

10. A variate can assume all values greater than zero and the equation of its probability density function is $kx\,e^{-x}$. Find k and also the mean of the distribution. Show that the median value is approximately 1·7.

11. A variate has the probability density function $kx^3(4 - x)^2$ $(0 \leqslant x \leqslant 4)$. Find k and also the mean and variance of the distribution.

12. The length of a side of a square is rectangularly distributed between 3 and 4. Show that its area A is distributed between 9 and 16 with a probability density function $\frac{1}{2} A^{-\frac{1}{2}}$. Sketch the frequency curve and calculate the mean and variance of the area of the square.

13. A radio valve has a life of x hours. If x is distributed according

to the probability density function $200/x^2$ $(x \geqslant 200)$ find the probability that neither of two such valves will have to be replaced during the first 300 h of operation.

14. A variate x is distributed according to the probability density function $kx\,e^{-\frac{x}{a}}$ $(x \geqslant 0, a$ positive) find: (a) The value of k in terms of a, (b) $\mathscr{E}(x)$, (c) $\mathscr{E}(x^2)$.

15. A plane carrying two bombs flies directly above a straight road. When a bomb is dropped its deviation from the road, measured perpendicularly to the road, is distributed with a probability density distribution of:

$$h(x) = \frac{50 + x}{2500} \qquad (-50 \leqslant x \leqslant 0)$$

$$= \frac{50 - x}{2500} \qquad (0 \leqslant x \leqslant 50)$$

(x is taken as positive or negative on either side of the road). If a bomb falls within 20 ft of the road the road will be damaged. If both bombs are dropped what is the probability that the road is damaged?

16. A population has a probability density function given by:

$$f(x) = k\,e^{1-(x/\theta)} \qquad (0 \leqslant x \leqslant \theta)$$

Find k and the mean value of this distribution. (Liv. U. part)

17. A variate x can assume values only between 0 and 1, and the equation of its frequency curve is

$$y = A\,e^{-2x} \qquad (0 < x < 1)$$

where A is a constant such that the area under the curve is unity. Determine the value of A to three decimal places. Calculate the mean and variance of the distribution. (J.M.B.)

18. The variable x has the continuous frequency function

$$f(x) = \frac{2}{\pi(1 + x^2/3)^2 \sqrt{3}} \qquad (-\infty < x < \infty)$$

Find the mean and standard deviation of the distribution of x.

(b) PQ is a fixed diameter of a circle of radius a. From P a variable chord is drawn, making an (acute) angle with PQ such that the probability that the angle lies between θ and $\theta + \delta\theta$ is $2\delta\theta/\pi$ over the range $0 \leqslant \theta \leqslant \frac{1}{2}\pi$. Find the distribution of the length of the chord, and show that the mean length is $4a/\pi$. (J.M.B.)

8

THE NORMAL DISTRIBUTION

8.1. INTRODUCTION

The *normal distribution* is probably the most widely studied distribution in statistics because:

(*a*) Distributions which are approximately normal are frequently encountered, e.g. most sets of random errors follow the normal distribution.

(*b*) The normal distribution is important as a 'limiting distribution', i.e. it can be used as an approximation to other distributions (see sections 8.7 and 8.8).

(*c*) The normal distribution is easy to use.

(*d*) It has been shown that the results obtained by assuming a non-normal population to be normally distributed are reasonably accurate when the departure from normality is not too severe.

(*e*) The central limit theorem shows that the means of samples of size n from any population are approximately normally distributed. The approximation improves as n gets bigger.

8.2. PROPERTIES OF THE NORMAL (OR GAUSSIAN) DISTRIBUTION

(*a*) The normal distribution is a continuous distribution unlike the Binomial and Poisson distributions which are discrete.

(*b*) The graph of the distribution is given by

$$y = f(x) = \frac{1}{\sigma\sqrt{(2\pi)}} \exp\left[-\frac{1}{2}\left(\frac{x-\mu}{\sigma}\right)^2\right]$$

where x can have all values from $-\infty$ to $+\infty$ and the parameters μ and σ^2 represent respectively the mean and variance of the distribution.

(*c*) The graph is bell-shaped and symmetrical about the point $x = \mu$, as shown in *Figure 8.1*.

The spread of the curve about μ depends on σ.

(*d*) It is a probability density function since

$$\int_{-\infty}^{+\infty} \frac{1}{\sigma\sqrt{(2\pi)}} \exp\left[-\frac{1}{2}\left(\frac{x-\mu}{\sigma}\right)^2\right] dx = 1$$

122

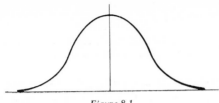

Figure 8.1

(*e*) Since it is a continuous probability density function, the probability that x lies between two specified values a and b is given by integration,

i.e. $\Pr(a \leqslant x \leqslant b) = \displaystyle\int_a^b \frac{1}{\sigma\sqrt{(2\pi)}} \exp\left[-\frac{1}{2}\left(\frac{x-\mu}{\sigma}\right)^2\right] dx.$

Note: read $\Pr(a \leqslant x \leqslant b)$ as the probability that x lies between a and b.

This is a difficult integral to evaluate. However, we can simplify it by carrying out the transformation

$$z = \frac{x-\mu}{\sigma} \qquad \dots (8.1)$$

Hence $dz = \dfrac{dx}{\sigma}$ and the integral becomes

$$\int_{z_1}^{z_2} \frac{1}{\sqrt{(2\pi)}} e^{-z^2/2} dz \qquad \dots (8.2)$$

where $\qquad z_1 = \dfrac{a-\mu}{\sigma}, \text{ and } z_2 = \dfrac{b-\mu}{\sigma}.$

This integral is independent of μ and σ except in so far as the limits depend on them.

From the laws of integration we have that

$$\int_{z_1}^{z_2} f(z)\,dz = \int_{-\infty}^{z_2} f(z)\,dz - \int_{-\infty}^{z_1} f(z)\,dz$$

and thus a table giving $\displaystyle\int_{-\infty}^{z} f(z)\,dz$, where z goes from $-\infty$ to $+\infty$ will give the values of all integrals of the type $\displaystyle\int_{z_1}^{z_2} f(z)\,dz.$

(*f*) Since the normal distribution is symmetrical, tables are given only for positive values of z. Most tables do not go above $z = 4$ since 99·99 per cent of the area under the normal curve lies within the range $z = -4$ to $z = +4$ and rarely can we hope to be accurate to within

0·01 per cent. The integral $\int_{-\infty}^{z_1} \frac{1}{\sqrt{(2\pi)}} e^{-z^2/2} \, dz$ gives the probability that z is less than z_1, i.e. $\Pr(z \leqslant z_1)$. Tables of this integral for $z_1 = 0$ to $z_1 = 4$ are given at the back of the book.

Example 8.1. Bolts coming off a production line are normally distributed with $\mu = 2.5$ in and $\sigma = 0.1$ in. Find the probability that a bolt chosen at random will lie between 2·6 and 2·7 in.

We are required to find

$$\Pr(2.6 \leqslant x \leqslant 2.7) = \int_{2.6}^{2.7} \frac{1}{0.1\sqrt{(2\pi)}} \exp\left[-\frac{1}{2}\left(\frac{x - 2.5}{0.1}\right)^2 \right] dx$$

In this case the transformation (8.1) is $z = \dfrac{x - 2.5}{0.1}$ which gives limits of integration for the transformed integral as

$$z_2 = \frac{2.7 - 2.5}{0.1} = 2, \text{ and } z_1 = \frac{2.6 - 2.5}{0.1} = 1$$

$$\begin{aligned}
\therefore \Pr(2.6 \leqslant x \leqslant 2.7) &= \int_{1}^{2} \frac{1}{\sqrt{(2\pi)}} e^{-z^2/2} \, dz \\
&= \Pr(z_1 \leqslant 2) - \Pr(z_2 \leqslant 1) \\
&= 0.9772 - 0.8413 \text{ (using tables)} \\
&= 0.1359.
\end{aligned}$$

(g) The curve given by $y = f(z) = [1/\sqrt{(2\pi)}] e^{-z^2/2}$ is called the *standardized normal curve* and $\dot{z} = (x - \mu)/\sigma$ is called the *standardized variate*.

(h) A shorthand way of writing 'is distributed normally with mean μ and variance σ^2' is 'is distributed $N(\mu, \sigma^2)$'. Thus the standardized normal distribution is $N(0,1)$.

8.3. USE OF NORMAL TABLES

Example 8.2. Find the probability that $z \leqslant 1.0$ (see *Figure 8.2*).

$$\Pr(z \leqslant 1.0) = \int_{-\infty}^{1} \frac{1}{\sqrt{(2\pi)}} e^{-z^2/2} \, dz = 0.8413$$

It can be seen from Examples 8.1 and 8.2 that only the standardized variate z is required when reading from the table. Also that the inclusion of the integral is superfluous and it will be omitted in future problems.

Since the total area under the curve is equal to 1 we can find $\Pr(z \geqslant z_1)$ by taking $1 - \Pr(z \leqslant z_1)$.

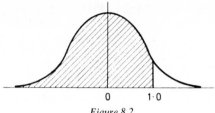

Figure 8.2

Example 8.3. Find the probability that $z \geqslant 0.04$.

$$\Pr(z \geqslant 0.04) = 1 - \Pr(z \leqslant 0.04) = 1 - 0.5160 = 0.484$$

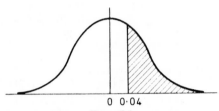

Figure 8.3

The symmetrical property of the normal distribution allows us to find probabilities for values of $z < 0$, as follows

$$\Pr(z \leqslant -z_1) = \Pr(z \geqslant z_1) = 1 - \Pr(z \leqslant z_1)$$

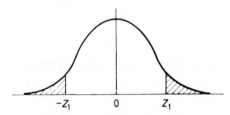

Figure 8.4

Example 8.4. $\Pr(z \leqslant -0.25) = \Pr(z \geqslant 0.25)$
$$= 1 - \Pr(z \leqslant 0.25)$$
$$= 1 - 0.5987$$
$$= 0.4013$$
Also $\Pr(z \geqslant -z_1) = \Pr(z \leqslant z_1)$

Example 8.5. $\Pr(z \geqslant -0.46) = \Pr(z \leqslant 0.46) = 0.6772$

We can find $\Pr(z_1 \leqslant z \leqslant z_2)$ using $\Pr(z \leqslant z_2) - \Pr(z \leqslant z_1)$

Example 8.6.
$$\begin{aligned}
\Pr(0{\cdot}5 \leqslant z \leqslant 2{\cdot}5) &= \Pr(z \leqslant 2{\cdot}5) - \Pr(z \leqslant 0{\cdot}5) \\
&= 0{\cdot}9938 - 0{\cdot}6915 \\
&= 0{\cdot}3023
\end{aligned}$$

Example 8.7.
$$\begin{aligned}
\Pr(-3{\cdot}75 \leqslant z \leqslant -1{\cdot}4) &= \Pr(1{\cdot}4 \leqslant z \leqslant 3{\cdot}75) \\
&= \Pr(z \leqslant 3{\cdot}75) - \Pr(z \leqslant 1{\cdot}4) \\
&= 0{\cdot}9999 - 0{\cdot}9192 \\
&= 0{\cdot}0807
\end{aligned}$$

Example 8.8.
$$\begin{aligned}
\Pr(-2{\cdot}4 \leqslant z \leqslant 1{\cdot}61) &\, \Pr(z \leqslant 1{\cdot}61) - \Pr(z \leqslant -2{\cdot}4) \\
&= \Pr(z \leqslant 1{\cdot}61) \\
&\qquad\qquad - [1 - \Pr(z \leqslant 2{\cdot}4)] \\
&= \Pr(z \leqslant 1{\cdot}61) + \Pr(z \leqslant 2{\cdot}4) - 1 \\
&= 0{\cdot}9463 + 0{\cdot}9918 - 1 \\
&= 0{\cdot}9381
\end{aligned}$$

Exercises 8a

Using tables of the normal distribution find:

1. (*a*) $\Pr(z \leqslant 2{\cdot}74)$, (*b*) $\Pr(z \leqslant 3{\cdot}09)$, (*c*) $\Pr(z \leqslant 0{\cdot}71)$.
2. (*a*) $\Pr(z \geqslant 1{\cdot}73)$, (*b*) $\Pr(z \geqslant 2{\cdot}65)$, (*c*) $\Pr(z \geqslant 3{\cdot}1)$.
3. (*a*) $\Pr(z \leqslant -1{\cdot}6)$, (*b*) $\Pr(z \leqslant -3{\cdot}5)$, (*c*) $\Pr(z \leqslant -2{\cdot}54)$.
4. (*a*) $\Pr(z \geqslant -2{\cdot}01)$, (*b*) $\Pr(z \geqslant -1{\cdot}4)$, (*c*) $\Pr(z \geqslant -3{\cdot}8)$.
5. (*a*) $\Pr(0{\cdot}24 \leqslant z \leqslant 0{\cdot}7)$, (*b*) $\Pr(1{\cdot}41 \leqslant z \leqslant 2{\cdot}5)$, (*c*) $\Pr(0{\cdot}31 \leqslant z \leqslant 1{\cdot}81)$.
6. (*a*) $\Pr(-2{\cdot}7 \leqslant z \leqslant 0{\cdot}03)$. (*b*) $\Pr(-0{\cdot}65 \leqslant z \leqslant 1{\cdot}96)$, (*c*) $\Pr(-1{\cdot}42 \leqslant z \leqslant 0)$.
7. (*a*) $\Pr(-0{\cdot}45 \leqslant z \leqslant -0{\cdot}11)$, (*b*) $\Pr(-2{\cdot}5 \leqslant z \leqslant -1{\cdot}41)$, (*c*) $\Pr(-1{\cdot}85 \leqslant z \leqslant -0{\cdot}6)$.

8.4. PRACTICAL PROBLEMS

When using the normal distribution to solve practical problems it must be remembered that the variate x can only be measured to a certain degree of accuracy. Since the normal distribution is continuous we must make allowances for the rounding off when choosing our standardized variate z. For example, if packets are weighed and the weights recorded to the nearest pound and we wish to find the probability that a given bag will weigh between 50 and 60 lb. Given that the weights are normally distributed mean 50 and variance 25, we must form our standardized variates as

$$z_2 = \frac{60{\cdot}5 - 50}{5} = 2{\cdot}1, \text{ and } z_1 = \frac{49{\cdot}5 - 50}{5} = -0{\cdot}1$$

The values 49·5 and 60·5 are used since we assume rounding off to the nearest pound is done by rounding up any fraction greater than 0·5 and rounding down any fraction less than 0·5 (any fraction equal to 0·5 is rounded to the nearest even integer). Thus 51·6 is taken as 52 and 51·3 is taken as 51.

Example 8.9. Assume the length of life of electric bulbs is normally distributed with mean 2000 h and standard deviation 250 h, the measurements being taken to the nearest hour. Find the probability that a given bulb will have a life between 1800 h and 2015 h.

$$z_1 = \frac{1799·5 - 2000}{250} = -\frac{200·5}{250} = -0·802$$

$$z_2 = \frac{2015·5 - 2000}{250} = \frac{15·5}{250} = 0·062$$

$$\Pr(-0·802 \leqslant z \leqslant 0·062) = \Pr(z \leqslant 0·802) + \Pr(z \leqslant 0·062) - 1$$
$$= 0·7887 + 0·5247 - 1$$
$$= 0·3134$$

Thus the probability of life between 1800 and 2015 h is 0·3134.

Example 8.10. In a sample of 100 screws taken from a population distributed normally $N(1·25, 0·0001)$ [see section 8.2 (*h*)] find the expected number of screws in the range 1·24 to 1·265.

$$z_1 = \frac{1·24 - 1·25}{0·01} = -1·0, z_2 = \frac{1·265 - 1·25}{0·01} = 1·5$$

$$\Pr(-1·0 \leqslant z \leqslant 1·5) = 0·8413 + 0·9332 - 1 = 0·7745$$

Expected number $= 100 \times 0·7745 = 77$ screws

Example 8.11. There are 175 men in a sample of size 1000 who have heights (measured to the nearest half inch) between the population mean μ and 5 ft $10\frac{1}{2}$ in. Given that the population is $N(\mu, 4)$ find μ.

The estimate of $\Pr(0 \leqslant z \leqslant z_2)$ is $\dfrac{175}{1000}$, i.e. 0·175.

$$\therefore \Pr(z \leqslant z_2) = 0·175 + 0·5 = 0·675$$

$$\therefore z_2 = 0·454 \qquad \qquad \dots\text{(i)}$$

$$\text{Now } z_2 = \frac{70·5 - \mu}{2} \qquad \qquad \dots\text{(ii)}$$

From (i) and (ii) $\dfrac{70\cdot5 - \mu}{2} = 0\cdot454$

∴ $\mu = 69\cdot592$

            $= 69\cdot5$ (to the nearest $\frac{1}{2}$ in)

8.5. THE USE OF STANDARDIZED VARIATE TO COMPARE THE RELATIVE MERITS OF VARIATES FROM DIFFERENT NORMAL DISTRIBUTIONS

Example 8.12. A student obtains marks as follows: 60 in Physics, 80 in Mathematics, 45 in English, and 58 in French. The marks are all distributed normally as follows: Physics $N(55,225)$, Mathematics $N(85,100)$, English $N(45,100)$ and French $N(56,64)$. Put the subjects in relative order of merit.

$$z_p = \frac{60 - 55}{15} = \frac{1}{3}, z_m = \frac{80 - 85}{10} = -\frac{1}{2}$$

$$z_e = \frac{45 - 45}{10} = 0, z_f = \frac{58 - 56}{8} = \frac{1}{4}$$

The order of merit is: Physics, French, English, Mathematics.

8.6. ARITHMETICAL PROBABILITY GRAPH PAPER

This type of graph paper has one scale which is uniform and one scale which is graduated according to the cumulative area under the normal curve. If the cumulative frequency polygon of a sample, size n, from a normal population is plotted on this graph paper the resulting graph should be a straight line.

Note: if a transformation has been carried out to make the distribution normal, then the transformed variate must be plotted on the uniform scale.

For $n > 50$ a fairly good straight line is obtained over the middle part of the curve. As an example (see *Figures 8.5* and *8.6*) samples of size 50 and 100 were selected at random from the normal tables and plotted. They give an indication as to how well we would expect the points to fit a straight line. Note that it is impossible to plot the extreme points 0 per cent and 100 per cent on the graph.

We thus have a method of testing if a sample comes from a normal population. It will also give us an approximation for the mean and the standard deviation of the (normal) parent population.

The population mean μ can be read off from the straight line at the point corresponding to 50 per cent. The standard deviation σ is calculated using the 95 and 5 per cent points, because $2 \times 1.645\sigma$ lies between these values

$$\sigma = \frac{z_{95\%} - z_{5\%}}{2 \times 1.645}$$

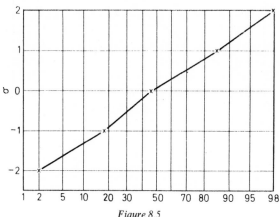

Figure 8.5

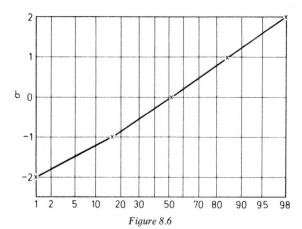

Figure 8.6

Example 8.13. Given the frequency distribution in *Table 8.1*, test if it is a sample from a normal population and estimate μ and σ^2. From *Figure 8.7* $\mu = 9.15$,

$$\sigma = \frac{10.73 - 7.55}{2 \times 1.645} = \frac{3.18}{3.29} = 0.97$$

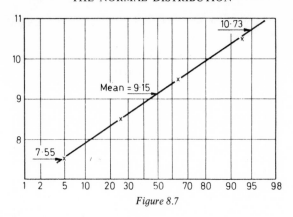

Figure 8.7

Table 8.1

Class interval	Frequency	Cumulative frequency	Cumulative relative percentage frequency
6·5–7·5	10	10	5
–8·5	36	46	23
–9·5	82	128	64
–10·5	56	184	92
–11·5	16	200	100

Exercises 8b
1. The frequency *Table 8.2* is thought to be a sample from a normal population. By plotting the cumulative relative percentage frequencies on arithmetic probability paper find approximations to μ and σ.

Table 8.2

Centre of interval	10	15	20	25	30	35	40	45	50	55
Frequency	2	5	6	14	26	18	13	10	3	3

2. Test to see if the data in *Table 8.3* is a sample from a normal distribution.

Table 8.3

x	34·5–	36·5–	38·5–	40·5–	42·5–	44·5–	46·5–	48·5–	50·5–
f	2	2	5	6	7	24	36	48	64

x	52·5–	54·5–	56·5–	58·5–	60·5–	62·5–	64·5–	66·5–68·5	
f	51	41	32	24	12	5	4	1	

8.7. THE NORMAL APPROXIMATION TO THE BINOMIAL DISTRIBUTION

For large n and not too small p, i.e. $np \geqslant 5$, the normal distribution can be used as an approximation to the binomial distribution. The mean and variance of the binomial distribution are np and npq and these are used as μ and σ^2 in the corresponding normal distribution.

Theorem

If x is a random variable which follows the binomial distribution then $\dfrac{x - np}{(npq)^{\frac{1}{2}}}$ has a distribution which approaches the standardized normal distribution as $n \to \infty$.

The binomial distribution is discrete whereas the normal distribution is continuous and so adjustments to the limits for classes have to be made as illustrated in the following example.

Example 8.14. Find, using the normal approximation to the binomial distribution, the probability of throwing 12 heads in 16 tosses of a fair coin.

$$\text{Fair coin implies } p = \frac{1}{2}$$

$$\therefore np = 16 \times \frac{1}{2} = 8, \, npq = 16 \times \frac{1}{2} \times \frac{1}{2} = 4.$$

To allow for the change from discrete to continuous variable, we allocate half of the probability between 11 and 12 and half that between 12 and 13 to 12. Thus we need to find the probability of getting between $11\frac{1}{2}$ and $12\frac{1}{2}$ heads.

$$\Pr(11\tfrac{1}{2} \leqslant x \leqslant 12\tfrac{1}{2}) = \Pr\left(z \leqslant \frac{12\tfrac{1}{2} - 8}{2}\right) - \Pr\left(z \leqslant \frac{11\tfrac{1}{2} - 8}{2}\right)$$

$$= \Pr(z \leqslant 2\cdot 25) - \Pr(z \leqslant 1\cdot 75)$$
$$= 0\cdot 9878 - 0\cdot 9599$$
$$= 0\cdot 0279$$

Using the binomial distribution the result would have been

$$\text{Pr}(12 \text{ Heads}) = {}^{16}C_{12}(\tfrac{1}{2})^{12} (\tfrac{1}{2})^4 = 0.0278.$$

From which it can be seen that the approximation compares favourably with the correct value.

Example 8.15. Find the probability of throwing between 10 and 15 ones (inclusive) in 48 tosses of a fair die.

$$\text{Pr}(1) = \frac{1}{6} \qquad \therefore np = 48 \times \frac{1}{6} = 8$$

$$npq = 48 \times \frac{1}{6} \times \frac{5}{6} = \frac{20}{3}$$

The range required using the normal approximation is 9.5 to 15.5. Thus the required probability is

$$\text{Pr}(9.5 \leqslant x \leqslant 15.5) = \text{Pr}\left(z \leqslant \frac{15.5 - 8}{\sqrt{(20/3)}}\right) - \text{Pr}\left(z \leqslant \frac{9.5 - 8}{\sqrt{(20/3)}}\right)$$

$$= \text{Pr}(z \leqslant 2.905) - \text{Pr}(z \leqslant 0.581)$$
$$= 0.9982 - 0.7193$$
$$= 0.2789$$

Example 8.16. The average number of defectives of a product which is being manufactured is 5 per cent. Find the probability that a lot of size 100 will contain (*a*) at least 9 defectives (*b*) 2 or 3 defectives.

$$(a)\ p = \frac{1}{20},\ np = \frac{100}{20} = 5,\ npq = 100 \times \frac{1}{20} \times \frac{19}{20} = \frac{19}{4}$$

The probability of at least 9 defectives is

$$\text{Pr}(x \geqslant 8.5) = \text{Pr}\left(z \geqslant \frac{8.5 - 5}{\sqrt{(19/4)}}\right) = \text{Pr}\left(z \geqslant \frac{7\sqrt{19}}{19}\right)$$

$$= \text{Pr}(z \geqslant 1.61) \qquad = 1 - \text{Pr}(z \leqslant 1.61)$$
$$= 1 - 0.9463 \qquad = 0.0537$$

(*b*) The probability of 2 or 3 defectives is

$$\text{Pr}(1.5 \leqslant x \leqslant 3.5) = \text{Pr}\left(z \leqslant \frac{3.5 - 5}{\sqrt{(19/4)}}\right) - \text{Pr}\left(z \leqslant \frac{1.5 - 5}{\sqrt{(19/4)}}\right)$$

$$= \text{Pr}\left(z \leqslant \frac{-3\sqrt{19}}{19}\right) - \text{Pr}\left(z \leqslant \frac{-7\sqrt{19}}{19}\right)$$

$$= \text{Pr}\left(z \leqslant \frac{7\sqrt{19}}{19}\right) - \text{Pr}\left(z \leqslant \frac{3\sqrt{19}}{19}\right)$$

$$= \Pr(z \leqslant 1{\cdot}61) - \Pr(z \leqslant 0{\cdot}69)$$
$$= 0{\cdot}9463 - 0{\cdot}7549$$
$$= 0{\cdot}1914$$

8.8. THE NORMAL APPROXIMATION TO THE POISSON DISTRIBUTION

For $a > 25$ the normal distribution can be used as an approximation to the Poisson distribution. The mean and the variance of the Poisson distribution are both equal to a and thus $\mu = a$, and $\sigma^2 = a$ for the corresponding normal distribution.

Theorem

If x is a random variate which follows the Poisson distribution then $(x - a)/\sqrt{a}$ has a distribution which approaches the standardized normal distribution as $a \to \infty$.

Example 8.17. A radioactive disintegration gives counts that follow a Poisson distribution. If the mean number of particles recorded in a 1 sec interval is 69, evaluate the probability of (a) less than 60 particles for a 1 sec interval, (b) more than 150 particles in a 2 sec interval, (c) more than 700 particles in a 10 sec interval.

(a) $$a = 69, z = \frac{59{\cdot}5 - 69}{\sqrt{69}} = \frac{-9{\cdot}5}{\sqrt{69}} = -1{\cdot}144$$

$$\Pr(z < -1{\cdot}144) = 1 - \Pr(z < 1{\cdot}144) = 1 - 0{\cdot}8737 = 0{\cdot}1263$$

(b) Because of the additive property of the Poisson distribution (see section 4.5) the counts in a 2 sec interval follow a Poisson distribution with mean $2 \times 69 = 138$.

$$z = \frac{150{\cdot}5 - 138}{\sqrt{138}} = \frac{12{\cdot}5}{\sqrt{138}} = 1{\cdot}064$$

$$\Pr(z > 1{\cdot}064) = 1 - \Pr(z < 1{\cdot}064) = 1 - 0{\cdot}8563 = 0{\cdot}1437.$$

(c) The count in a 10 sec interval has mean of 690 and is a Poisson distribution.

$$\therefore z = \frac{700{\cdot}5 - 690}{\sqrt{690}} = \frac{10{\cdot}5}{\sqrt{690}} = 0{\cdot}4$$

$$\Pr(z > 0{\cdot}4) = 1 - \Pr(z < 0{\cdot}4) = 1 - 0{\cdot}6554 = 0{\cdot}3446.$$

EXERCISES 8
1. Find the area under the normal curve
(a) between $z = -1\cdot31$ and $z = 2\cdot43$,
(b) between $z = 1\cdot23$ and $z = 1\cdot87$,
(c) between $z = -2\cdot64$ and $z = -0\cdot47$,
(d) to the left of $z = -1\cdot65$,
(e) to the right of $z = 2\cdot06$,
(f) corresponding to $z \geqslant 2\cdot16$,
(g) corresponding to $-0\cdot4 \leqslant z \leqslant 1\cdot83$,
(h) outside the range $z = -2\cdot41$ to $z = 1\cdot68$.

2. If z is normally distributed with mean 0 and variance 1 find: (a) $\Pr(z \leqslant -1\cdot92)$, (b) $\Pr(-1\cdot96 \leqslant z \leqslant 1\cdot96)$, (c) $\Pr(z \geqslant 1)$, (d) $\Pr(-0\cdot675 \leqslant z \leqslant 0\cdot675)$.

3. For the standardized normal curve, find the value of z_1 such that: (a) the area to the right of z_1 is $0\cdot2266$, (b) the area to the left of z_1 is $0\cdot0217$, (c) the area between $z = -0\cdot23$ and $z = z_1$ is $0\cdot5642$, (d) the area between $z = 0\cdot51$ and $z = z_1$ is $0\cdot1627$.

4. The weights of students are thought to be normally distributed with mean weight 11 st. 8 lb and standard deviation 1 st. 8 lb. Find: (a) The expected number of students in a sample of size 150 who have weights greater than 13 st. 6 lb, (b) the range, equally spaced about the mean, that contains 60 per cent of the weights. Assume that the weights are rounded off to the nearest pound.

5. If a variate is distributed $N(9\cdot2, 1\cdot69)$ find the probability that a single variate selected at random will be (a) greater than or equal to 12, (b) less than or equal to 7.

6. The cost per packet of biscuits is normally distributed mean 1s. 3d. and standard deviation 3d. What percentage of packets cost between 11d. and 1s. 6d.?

7. If the examination marks in Mathematics are normally distributed mean 70 and variance 100, and the marks are graded in classes 40–49 grade A, 50–59 grade B, etc., find how many students in a class of 400 obtained each of the grades A, B, C, D, E, and F.

8. In an examination the marks were normally distributed mean 70 and variance 100. If 14 students obtained Grade D, how many students sat for the examination? (Use the same gradings as in question 7.)

9. In a sample from a normal distribution, mean $14\cdot0$ and standard deviation $3\cdot3$, 253 variates have values exceeding $15\cdot3$. How large is the sample?

10. If 10 per cent of light bulbs have lives exceeding 2340 h and the mean life is 2200 h. Find: (a) the standard deviation of the distribution if it is assumed to be normal, (b) what percentage of bulbs have lives between 2150 and 2410 h, (c) what range, equally spaced about the mean, contains 75 per cent of the distribution?

11. Compare the merits of the following variates obtained from normal distributions (a) 15·4 from $N(12·0.$ 4·00), (b) 29·6 from $N(30, 9·00)$, (c) 27·8 from $N(28, 1·00)$.

12. Packages from a packaging machine have a gross weight which is normally distributed mean 1·01 lb and variance 0·01 lb. What is the probability that a package weighs less than 1·00 lb?

13. The average mark for an examination was 70 and the standard deviation was 10. Twenty-four students had marks in the range 61–79. If the marks follow a normal distribution how many students took the examination?

14. In a binomial distribution the probability of a success was $p = \frac{1}{4}$, find the probability of obtaining: (a) 22 successes in 80 trials, (b) between 14 and 18 successes (inclusive) in 96 trials.

15. An unbiased coin is tossed 200 times. Find the probability that (a) exactly 99 heads occur, (b) 116 or more heads occur.

16. Find the probability of getting, in 96 tosses of an unbiased die (a) exactly 16 ones, (b) at least 60 even results, (c) less than 40 odd results.

17. A pair of unbiased dice are rolled 100 times. What is the probability that a 6, 7, 8, or 9 will occur 65 times?

18. According to the Mortality Tables the probability of a man dying while he is 64 is 0·11. If 10,000 people have life insurance policies in force on their 64th birthday find the probability that more than 1175 of them will die before a year has gone by.

19. The number of telephone calls to an exchange has an average value of 16 per minute. Find the probability that more than 22 calls will come into the exchange in any given minute.

20. Out of 1000 patients, 950 responded to a certain drug. What is the probability that out of a further 100 patients 10 do not respond.

21. By using the normal approximation to the binomial distribution find the number of throws of a fair die necessary to ensure that the probability of obtaining at least 6 sixes is one half.

22. The disintegrations from a radioactive sample are being counted. The true average rate is determined to be 12 counts/sec. Using the normal approximation to the Poisson distribution determine the probability that in any 1 sec the number of counts will be (a) 19, 20 or 21, (b) greater than 14, (c) lie between 9 and 16 inclusive.

23. The data given in *Table 8.4* are the breaking strains in lb wt. to the nearest pound at different points in a length of manufactured cord. Prepare a frequency table, grouping the data in equal class intervals, one such interval being 1000 to 1004 inclusive. Obtain the estimates of the population mean breaking strength μ and standard deviation σ (estimates are $\bar{x}$ and s). Plot the relative cumulative frequencies on **normal probability paper and draw the best (according to your eye)**

F

Table 8.4

992	1007	1001	1010	1001	1008	1013	1010	995	1003	1009	990
1003	1008	999	1002	992	990	998	1013	1014	992	991	1006
998	1001	991	998	1011	985	986	996	1010	1008	1008	1024
995	1023	1001	998	998	1004	1010	1000	1002	997	1013	1005
1006	1015	994	1002	999	1008	1025	998	1005	1008	992	1016
1003	981	1005	982	997	1004	1013	1018	984	1004	1009	1001
1014	1018	991	997	1009	1003	998	1007	1028	1000	988	1012
1012	1001	979	992	1002	989	1010	1017	1011	996	1006	1009
994	1011	976	1003								

straight line through the points. From this straight line again estimate μ and σ.

24. In an examination the students were placed as follows: 10 per cent Grade A, 20 per cent Grade B, 47 per cent Grade C, 13 per cent Grade D, and 10 per cent Grade E. If Grade C ranges from 47 to 61 and the marks are normally distributed, calculate the mean and standard deviation of the distribution.

25. *Table 8.5* gives the Labour vote as a percentage of (Labour and Conservative) vote in 1945, in the 576 constituencies contested by both parties and won by one of them.
Draw a histogram of this distribution and calculate the corresponding

Table 8.5

Percentages	Number of constituencies
Below 35	41
35–	87
45–	126
55–	179
65–	97
over 75	46

expected frequencies for a normal distribution with mean 57·2 per cent and standard deviation 13·5 per cent. (Liv. U.)

26. State how the normal distribution may be used as an approximation to the binomial when the sample size is large.

Components are packed in boxes of 500, and from past experience the proportion defective is known to be 0·30. Calculate the probability that a box contains at least 160 defective components. Also, calculate the probability that in four boxes inspected, exactly two contain at least 160 defectives. (Liv. U.)

27. The Ruritanian Sugar Corporation owns a machine which weighs out and packs up packets of sugar, each of whose contents is nominally 1 lb. It is known from long experience that the actual weight of sugar in the packets is normally distributed and has a standard deviation of 0·01 lb. Periodically an inspector visits the factory and weighs the contents of 20 packets taken at random from the production. Show that if the risk that he finds at least one packet underweight is to be 1 in 1000, the factory must on average give about 4 per cent overweight. (Liv. U.)

28. (a) An insurance company finds that 0·005 per cent of the population dies from a certain kind of accident each year. Ten thousand people are insured with the company against this risk. Calculate (to two significant figures) the probability that the company will receive at least four claims for this accident in any given year.

(b) A die is thrown 300 times, a score of 1 or 2 being counted as a 'success'. Use the normal approximation to the binomial distribution to derive (to two decimal places) the probability that the number of 'successes' will not deviate from 100 by more than 15. (J.M.B.)

29. The following are the weights in ounces of a random sample of fifteen 2 lb packets of sugar delivered by an automatic packing machine: 32·11, 31·97, 32·18, 32·03, 32·25, 32·07, 32·05, 32·14, 32·19, 31·98, 32·07, 31·99, 32·16, 32·03, 32·18. Calculate the mean and standard deviation. Assuming the distribution to be normal, estimate the percentage of underweight packets which the machine is delivering.

The Board of Trade stipulates that such a machine must not deliver more than 5 per cent underweight packets. If the machine is to comply with this regulation and if its variability cannot be better controlled, calculate the value to which the mean must be raised. (J.M.B.)

30. Jackets for young men are made in the following sizes, according to chest measurement.

Size	1	2	3	4	5	6
Chest measurement (in)	30–	32–	34–	36–	38–	40–42

The chest measurements of young men in a certain age range are known to be normally distributed with mean 35·63 in and standard deviation 2·00 in. Estimate to the nearest unit the percentages of young men in this age range likely to require each of the six sizes and also the percentages likely to fall above or below the size range.

(J.M.B. part)

9

SIGNIFICANCE TESTING AND CONFIDENCE INTERVALS

9.1. INTRODUCTION

VERY often in statistics it is necessary to test hypotheses about the parameters of a population. A *statistical hypothesis* is an assumption made about some parameter. This assumption could be completely verified if the whole population could be examined. However, in most cases only estimates of the parameters obtained from random samples are available and the assumptions must be tested using these estimates. These tests are called *tests of significance* or *tests of hypotheses*.

The assumption about the parameters is called the *null hypothesis* and is generally denoted by H_0, while the alternative to the null hypothesis is called the *alternative hypothesis* and is generally denoted by H_1.

In this chapter we shall restrict ourselves to tests involving the normal or student's 't' distributions, i.e. we shall assume that the estimates of the parameters are normally distributed. If we standardize the parameter using the null hypothesis H_0, i.e.

$$z = \frac{\text{Estimate} - \text{Null hypothesis value of parameter}}{\sqrt{(\text{variance of the estimate})}} \quad \dots (9.1)$$

then z follows either the normal distribution $N(0, 1)$ or the 't' distribution. If the value of z occurs in the tail of the distribution then this can be due to one of two things, either a not very probable event has occurred or the original null hypothesis was wrong. In statistical testing we conclude that the second of these two possibilities is the correct one, with the added proviso that the probability of our conclusion being wrong is equal to the probability of the rare event mentioned in the first possibility. This probability fixes the extent of the tail of the distribution. It fixes critical values z_α outside which a value of z will give a *significant result*. A value of z inside these critical values gives a *non-significant result*.

A significant result means that we reject the null hypothesis in favour of the alternative hypothesis, while a non-significant result means we do not reject the null hypothesis in favour of the alternative hypothesis.

Note: a non-significant result does not mean that we accept the null

139

hypothesis. As will be seen from the section on confidence intervals the null hypothesis is one of many hypotheses which could be true, according to the data.

The symbol z_α was used for the critical values of z since $\alpha/100$ is the probability associated with obtaining a value outside the critical values. Using the normal distribution, taking α to be 5 per cent and considering $2\frac{1}{2}$ per cent at both ends of the distribution the critical values are $z = -1.96$ and $z = 1.96$ (see *Figure 9.1*).

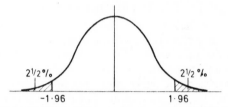

Figure 9.1. Two-tailed test using the normal distribution

If we obtain a value of z outside the range ± 1.96 we have a significant result at the 5 per cent level and reject the null hypothesis. It is more generally stated that 'the result is significant at the 5 per cent level'. Thus α is the level of significance and must be chosen before the test is carried out. It will depend on the consequences of rejecting a true null hypothesis.

If we are only interested in one side of a distribution the test is said to be *one-tailed*. *Figure 9.2* illustrates this for a normal distribution.

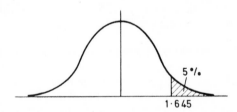

Figure 9.2. One-tailed test using the normal distribution

Figure 9.1 illustrates a *two-tailed* test. Whether it is a one-tailed or a two-tailed test depends on the alternative hypothesis. For example:

(*a*) The null hypothesis $\mu = \mu_0$ and the alternative hypothesis $\mu \neq \mu_0$ entails a two-tailed test.

(*b*) The null hypothesis $\mu = \mu_0$ and the alternative hypothesis $\mu > \mu_0$ entails a single-sided test.

A two-tailed test is carried out when the experimenter is interested in whether or not the null hypothesis value has changed. He is not interested in whether the change is an increase or a decrease.

A single-tailed test is carried out when only changes in a given direction (increase or decrease) are of interest.

The problem 'test if a coin is biased' would require a two-tailed test since the coin could be biased towards either heads or tails.

The problem 'test if a coin is biased in favour of heads' would require only a single-sided test.

Example 9.1. A new process is introduced into a factory. Is a single-tailed or two-tailed test required to test if the process produces a superior product?

Single-tailed test, since a test for a worse product is not required.

Example 9.2. Test at the 5 per cent level if the single sample value 15·4 comes from a normal population mean $\mu = 15·0$ and known variance 0·09.

The null hypothesis is $\mu = 15·0$ and the alternative hypothesis is $\mu \neq 15·0$.

$$\text{Standardizing, we have } z = \frac{15·4 - 15·0}{0·3} = 1·33^{\cdot}$$

Since we have a two-tailed test we test if z is between $\pm 1·96$. This is equivalent to testing $|z| < 1·96$.

$1·33^{\cdot} < 1·96$ and is not significant, so that we cannot reject the null hypothesis.

Example 9.3. From a normal population whose variance is 4 a member is selected at random and found to be 16. Test at the 1 per cent level of significance the null hypothesis that the population mean μ is 21, against the alternative hypotheses: (a) $\mu \neq 21$. (b) $\mu < 21$.

(a) $z = (16 - 21)/2 = -2·5$, $\therefore |z| = 2·5$, $z_{1\%} = 2·58$, $2·5 < 2·58$ and is not significant, therefore we cannot reject the null hypothesis at the 1 per cent level in favour of the alternative $\mu \neq 21$ [see *Figure 9.3(a)*].

(b) $z_{1\%} = 2·33$, $2·5 > 2·33$ and is significant, therefore we reject the null hypothesis at the 1 per cent level in favour of the alternative $\mu < 21$ [see *Figure 9.3(b)*].

Exercises 9a

1. A single random sample, value 149·7, was taken from a population whose standard deviation is 10·4. Test the null hypothesis that the population mean μ is 131 against the alternative hypotheses: (a) $\mu \neq 131$, (b) $\mu > 131$. Use both 5 per cent and 1 per cent levels of significance.

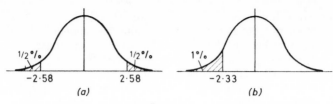

Figure 9.3

2. In a given interval of time 125 particles of radiation were counted. Assuming that the number of particles in a given interval follow the Poisson distribution, test the hypothesis that the mean number of particles emitted in the given interval is 100 against the alternative hypothesis that the mean number is greater than 100. Use a 1 per cent level of significance.

3. A coin is tossed 100 times and the number of heads recorded is 60. Use the normal approximation to the binomial distribution to test if the coin is biased.

4. A person states that he can tell by taste whether milk has been poured into a cup before or after the tea. Out of 50 cups tested 30 are identified correctly. It is unknown to the taster how many of each kind are present. Are the results significant at the 5 per cent level?

9.2. TESTS OF A SAMPLE MEAN

So far we have only considered samples of size 1. In general we have a random sample size $n > 1$ and the best estimate of the population mean is given by the sample mean $\bar{x}$. Thus we must test our null hypothesis about the population mean using the sample mean. We make use of the following theorems:

Theorem 1

The mean of means of all possible samples of size n drawn from a population equals the mean of the population,

i.e. $\qquad \mathscr{E}(\bar{x}) = \mu \quad$ (see section 3.6) $\qquad \ldots (9.2)$

Theorem 2

The variance of the sampling distribution of means of samples of size n is equal to $1/n$ times the population variance,

i.e. $\qquad \mathrm{var}(\bar{x}) = \frac{1}{n} \mathrm{var}(x) = \frac{\sigma^2}{n} \quad$ (see Chapter 14) $\qquad \ldots (9.3)$

Theorem 3

The sampling distribution of means of samples of size n from a normal population is itself normal.

Theorem 4

For a large variety of non-normal populations the sampling distribution of means of samples of size n is approximately normal. The approximation improves as n increases.

Thus if we have a sample of size n we form $\bar{x}$ and using equations (9.2) and (9.3) we test the statistic

$$z = \frac{\bar{x} - \mu}{\sigma/\sqrt{n}} \qquad \dots (9.4)$$

against the normal tabular value.

Example 9.4. A sample of size 64 has a mean value $\bar{x} = 36\cdot5$. The population variance is 16. Test at the 1 per cent level the null hypothesis $\mu = 35\cdot3$ against (*a*) the alternative $\mu \neq 35\cdot3$, (*b*) the alternative $\mu > 35\cdot3$.

$$z = \frac{36\cdot5 - 35\cdot3}{4/\sqrt{64}} = \frac{1\cdot2}{4/8} = 2\cdot4$$

(*a*) $z_{1\%}$ (two-tailed) $= 2\cdot58$, $2\cdot4 < 2\cdot58$ therefore we cannot reject at the 1 per cent level the null hypothesis in favour of $\mu \neq 35\cdot3$.

(*b*) $z_{1\%}$ (one-tailed) $= 2\cdot33$, $2\cdot4 > 2\cdot33$ therefore we reject at the 1 per cent level the null hypothesis in favour of $\mu > 35\cdot3$.

Example 9.5. The mean weight of men in England is 170 lb with a standard deviation of 20 lb. A random sample of 100 men taken in Liverpool has a mean of 164 lb. Test at the 5 per cent level to see if the average weight of men in Liverpool is different from the overall country average.

This is a two-tailed test and we must test if $\bar{x} = 164$ comes from a population mean $\mu = 170$ and variance $\dfrac{\sigma^2}{n} = \dfrac{20^2}{100} = 4$.

$$z = \frac{164 - 170}{2} = -3$$

$$z_{5\%} = 1\cdot96, \; |z| = 3 > 1\cdot96$$

Thus there is evidence that the average weight of Liverpool men is different from that for the country as a whole.

Exercises 9b

1. The life of electric light bulbs produced by a company have a standard deviation of 120 h. The mean was 2500 h, but it is thought that this has now changed due to the introduction of new materials. A sample of 30 electric light bulbs have a mean life of 2548 h. Does this sample support the idea of a change? Take a 5 per cent level of significance.

2. If in the previous problem we were only interested in an increase in the mean would this have affected the result?

3. A car company claims that its cars do 100 miles on 3 gal of petrol, with a standard deviation of 0·2 gal. 18 cars are tested and they have a mean consumption of 3·15 gal per 100 miles. Does this result refute the company's claim?

9.3. DIFFERENCE OF TWO POPULATION MEANS

Suppose $\bar{x}_1$ and $\bar{x}_2$ are the means of two independent random samples of sizes n_1 and n_2 from populations with unknown means μ_1 and μ_2 and known variances σ_1^2 and σ_2^2 respectively. $\bar{x}_1 - \bar{x}_2$ is distributed approximately normally with a population mean $\mu_1 - \mu_2$ and variance $\sigma_1^2/n_1 + \sigma_2^2/n_2$.

Thus to test the hypothesis about the value of the difference $\mu_1 - \mu_2$ we form the standardized variable

$$z = \frac{\bar{x}_1 - \bar{x}_2 - (\mu_1 - \mu_2)}{\sqrt{(\sigma_1^2/n_1 + \sigma_2^2/n_2)}} \qquad \ldots (9.5)$$

and test against a value from the normal table.

Note: if $\sigma_1^2 = \sigma_2^2 = \sigma^2$ (known) then the standardized variable reduces to

$$z = \frac{\bar{x}_1 - \bar{x}_2 - (\mu_1 - \mu_2)}{\sigma\sqrt{(1/n_1 + 1/n_2)}} \qquad \ldots (9.6)$$

The most common hypothesis is that $\mu_1 = \mu_2$, in which case equations (9.5) and (9.6) reduce to

$$z = \frac{\bar{x}_1 - \bar{x}_2}{\sqrt{(\sigma_1^2/n_1 + \sigma_2^2/n_2)}} \qquad \ldots (9.7)$$

$$z = \frac{\bar{x}_1 - \bar{x}_2}{\sigma\sqrt{(1/n_1 + 1/n_2)}} \qquad \ldots (9.8)$$

Example 9.6. Two independent random samples of sizes 8 and 10 respectively are drawn from populations whose variances are 4 and 1 respectively. The means of the samples are 54·5 and 56·2. Is there evidence to suggest that there is no difference in the population means?

The null hypothesis is that $\mu_1 = \mu_2$ so we use 9·7.

$$z = \frac{56\cdot2 - 54\cdot5}{\sqrt{(\frac{4}{8} + \frac{1}{10})}} = \frac{1\cdot7}{\sqrt{0\cdot6}} = 2\cdot2$$

The alternative hypothesis is $\mu_1 \neq \mu_2$ and so we have a two-tailed test

$$z_{5\%} = 1\cdot96, \; z_{1\%} = 2\cdot58.$$

We conclude that at the 5 per cent level the evidence is against equal population means but at the 1 per cent level the evidence is in favour of equal population means.

Exercises 9c

1. For the data of Example 9.6 test the null hypothesis that: (a) $\mu_1 - \mu_2 = 1\cdot1$ against the alternative $\mu_1 - \mu_2 < 1\cdot1$, (b) $\mu_1 - \mu_2 = -2\cdot0$ against the alternative $\mu_1 - \mu_2 > -2\cdot0$.

2. In *Table 9.1* sample A is taken from a population with a variance of 3·6 while sample B is taken from a population with variance 12·4. Test if the population means are the same at the 5 per cent level of significance.

Table 9.1

Sample A					Sample B			
47·1,	48·9,	49·2,	48·6,	48·6,	50·1,	53·0,	62·2,	52·8,
47·5,	50·4,	47·3,	49·2,	49·2,	48·4,	49·4,	52·9,	48·9
49·3,	47·3,							

9.4. TEST FOR PAIRED DATA

Suppose we have two independent samples of equal sizes n from populations with means μ_1 and μ_2 and variances σ_1^2 and σ_2^2. If the values can be paired one from each sample we obtain one sample, size n, of paired values $d_i = x_{1,i} - x_{2,i}$ $(i = 1, 2, \ldots n)$, which have come from a population with mean $\mu_1 - \mu_2$ and variance $\sigma_1^2 + \sigma_2^2$. We can test a null hypothesis value $\bar{d}$, of $\mu_1 - \mu_2$ using the standardized variable

$$z = \frac{\bar{d} - (\mu_1 - \mu_2)}{\sqrt{[(\sigma_1^2 + \sigma_2^2)/n]}} \qquad \ldots (9.9)$$

The most common test value for $\mu_1 - \mu_2$ is zero and (9.9) reduces to

$$z = \frac{\bar{d}}{\sqrt{[(\sigma_1^2 + \sigma_2^2)/n]}} \qquad \ldots (9.10)$$

Example 9.7. Five pieces of cloth are cut in half and one half of each piece is treated with a strengthening compound. Both treated and untreated pieces are tensioned until they break. The weights required to break the pieces of cloth are given in the following table. From previous experiments of this type it is known that the variance of each population of possible weights is 0·5 lb. Test if the compound has had any effect. Use a 10 per cent level of significance.

Treated	20·5	21·3	19·8	18·4	20·6
Untreated	20·1	20·8	19·7	18·5	20·0
d	+0·4	+0·5	+0·1	−0·1	+0·6

$\Sigma d = 1·5$, $\bar{d} = 0·3$. Using equation (9.10)

$$z = \frac{0·3}{[(0·5 + 0·5)/5]} = 0·6708$$

This is a one-tailed test because if the null hypothesis is $\mu_t - \mu_u = 0$ the alternative hypothesis is $\mu_t > \mu_u$.

$z_{10\%}$ (single-sided) $= 1·28$ and $0·6708 < 1·28$, therefore there is no evidence that the compound has had any effect.

Exercises 9d
1. One bolt of each size from $\frac{1}{8}$ ($\frac{1}{8}$) 1 in is treated to strengthen it. The bolts are then tested for breaking strength. A similar set of untreated bolts is tested at the same time. The results are given in *Table 9.2*. Test if the treatment has had a significant effect. Use a 5 per cent level of significance.

Table 9.2

Size	$\frac{1}{8}$ in	$\frac{1}{4}$ in	$\frac{3}{8}$ in	$\frac{1}{2}$ in	$\frac{5}{8}$ in	$\frac{3}{4}$ in	$\frac{7}{8}$ in	1 in
Treated	15·6	25·1	34·7	43·3	55·6	61·2	70·4	82·4
Untreated	15·4	26·2	35·1	44·0	55·7	61·8	70·6	83·1

9.5. TEST FOR A POPULATION MEAN GIVEN A LARGE SAMPLE (POPULATION VARIANCE UNKNOWN)

So far we have been given the population variance σ^2. This is not generally the case. In most problems σ^2 is unknown and has to be estimated by the sample variance $s^2 = \Sigma(x - \bar{x})^2/(n - 1)$. If the sample size n is greater than 30 the tests of significance are carried out in the same way as when σ^2 was known. That is, s^2 replaces σ^2 in the standardized formulae.

Example 9.8. For a sample of size 40 the mean is $\bar{x} = 4\cdot65$ and the standard deviation is $s^2 = 0\cdot9$. Test at the 5 per cent level if the sample could have come from a population whose mean μ is $4\cdot5$.

$$z = \frac{\bar{x} - \mu}{s/\sqrt{n}} = \frac{4\cdot65 - 4\cdot5}{\sqrt{0\cdot9}/\sqrt{40}} = 0\cdot15 \sqrt{\left(\frac{400}{9}\right)} = 1\cdot0$$

$z_{5\%} = 1\cdot96$ (the alternative hypothesis is $\mu \neq 4\cdot5$ and so we use a two-tailed test).

The result is not significant and we conclude that the population mean could be $4\cdot5$.

Example 9.9. The mean lifetime of a sample of 100 fluorescent light bulbs is computed to be 1570 h with a standard deviation of 120 h. If μ is the mean lifetime of all bulbs produced, test the hypothesis that $\mu = 1600$ against the alternative that $\mu \neq 1600$

$$z = \frac{1570 - 1600}{120/\sqrt{100}} = -2\cdot5$$

$$z_{5\%} = 1\cdot96, \qquad z_{1\%} = 2\cdot58$$

Thus we reject the hypothesis $\mu = 1600$ in favour of the alternative hypothesis $\mu \neq 1600$ at the 5 per cent level, but we do not reject the hypothesis at the 1 per cent level.

Example 9.10. Table 9.3 is a sample from a population whose mean is thought to be $3\cdot5$. Test whether the mean is $3\cdot5$ or is less than $3\cdot5$.

Table 9.3

x	0	1	2	3	4	5	6
Frequency	4	9	16	34	40	13	5

$$\Sigma fx = 398, \qquad \Sigma fx^2 = 1524, \qquad \Sigma f = 121$$

$$\bar{x} = 3\cdot29, \qquad s^2 = \frac{1524 \times 121 - 398^2}{120 \times 121} = 1\cdot79$$

$$z = \frac{3\cdot29 - 3\cdot5}{\sqrt{(1\cdot79/121)}} = -1\cdot73$$

$$z_{5\%} = 1\cdot645, \qquad z_{1\%} = 2\cdot33 \text{ (single-tailed test)}$$

Thus we reject the null hypothesis at the 5 per cent level, but not at the 1 per cent level.

9.6. TESTS FOR THE DIFFERENCE BETWEEN TWO POPULATION MEANS GIVEN TWO LARGE SAMPLES (POPULATION VARIANCES UNKNOWN)

(a) The variances of the two populations are assumed equal, i.e.

$\sigma_1^2 = \sigma_2^2 = \sigma^2$ (This can also be tested using the 'F' distribution, see Chapter 12).

Two samples, one from each population, have means $\bar{x}_1$ and $\bar{x}_2$, variances s_1^2 and s_2^2 and are of sizes n_1 and n_2. To test if the population of all such differences $x_1 - x_2$ has a mean $\mu_1 - \mu_2$ we first find s^2 as an estimate of σ^2.

$$s^2 = \frac{(n_1 - 1)s_1^2 + (n_2 - 1)s_2^2}{n_1 + n_2 - 2} = \frac{\Sigma(x - \bar{x}_1)^2 + \Sigma(x - \bar{x}_2)^2}{n_1 + n_2 - 2}$$

We then form

$$z = \frac{\bar{x}_1 - \bar{x}_2 - (\mu_1 - \mu_2)}{s\sqrt{(1/n_1 + 1/n_2)}} \qquad \dots (9.11)$$

and test using the normal distribution.

Example 9.11. Two samples of electric light bulbs are treated for length of life. The results are given in *Table 9.4*.

Table 9.4

	Sample I	Sample II
Number in sample	42	92
Mean of sample ($= \bar{x}$)	2060	2040
$\Sigma(x - \bar{x})^2$	20,100	18,300

Are the samples from populations with different means (assuming equal population variances)?

$$s^2 = \frac{20,100 + 18,300}{42 + 92 - 2} = 291$$

The null hypothesis is $\mu_1 = \mu_2$

$$\therefore \qquad z = \frac{2060 - 2040}{\sqrt{[291(1/42 + 1/92)]}} = 20 \times \sqrt{\left(\frac{42 \times 46}{291 \times 67}\right)}$$

$$= 20 \times \sqrt{\left(\frac{1932}{19,497}\right)}$$

$$= 20 \times 0\cdot3147$$

$$= 6\cdot294 > 1\cdot96$$

Thus we reject the hypothesis $\mu_1 = \mu_2$ at the 5 per cent level (and also at the 1 per cent level).

(b) The population variances are unequal ($\sigma_1^2 \neq \sigma_2^2$).
So long as the samples are large we can use the statistic

$$z = \frac{\bar{x}_1 - \bar{x}_2 - (\mu_1 - \mu_2)}{\sqrt{(s_1^2/n_1 + s_2^2/n_2)}}$$

Example 9.12. Screws are made by two different machines A and B. Samples of size 120 and 150 are taken from the output of machines A and B respectively and in *Table 9.5* results of measuring the lengths of the screws are obtained.

Table 9.5

	A	B
Σx	492	690
$\Sigma(x - \bar{x})^2$	12	70

Test if the machines are set to give different lengths of screws.

$$\bar{x}_A = 4\cdot1, \qquad\qquad \bar{x}_B = 4\cdot6$$
$$s_A^2 = 12/119, \qquad\qquad s_B^2 = 70/149$$
$$z = \frac{4\cdot1 - 4\cdot6}{\sqrt{[12/(119 \times 120) + 70/(149 \times 150)]}} = -0\cdot5 \times \sqrt{253\cdot5}$$
$$= -3\cdot18$$

Since $z_{5\%} = 1\cdot96$ the conclusion is that the machines are set to give different lengths.

9.7. TESTS FOR POPULATION MEANS GIVEN SMALL SAMPLES (POPULATION VARIANCES UNKNOWN)

If σ^2 is unknown we must estimate it by s^2 as in sections 9.5 and 9.6. However, because of the small sample size we must replace the statistic z by the statistic

$$t = \frac{\bar{x} - \mu}{s/\sqrt{n}} \qquad \ldots (9.12)$$

This statistic does not follow the normal distribution for small n. It was first studied by W. S. Gossett, who was statistician to Guinness Brewery, and who wrote under the pen name of 'Student'. He developed a distribution which is now called 'Student's t distribution' and is followed by equation (9.12).

Definition

If $z = \dfrac{\bar{x} - \mu}{\sigma/\sqrt{n}}$ follows the normal distribution and $\chi^2 = \dfrac{vs^2}{\sigma^2}$ follows

the chi-squared distribution (see Chapter 11) then $t = \dfrac{z}{\sqrt{(\chi^2/v)}} =$

$\dfrac{\bar{x} - \mu}{s/\sqrt{n}}$ follows the t distribution.

A table of values of t for varying degrees of freedom v is given in *Table 5* (Appendix). This table is given for a two-tailed test, that with an alternative hypothesis of the form $\mu \neq \mu_0$. Note that for an infinite number of degrees of freedom the t test gives the same values as the normal distribution. $t_{\alpha\%}(v)$ means the value of t with v degrees of freedom at the $\alpha\%$ level of significance. v is the denominator used in the calculation of s^2.

Example 9.13. The following are a random sample of size 4: 12·1, 12·4, 12·5, 12·8. Test if the population mean is 12·0. Use a 5 per cent level of significance.

Using a false origin of 12·0 the sample is 0·1, 0·4, 0·5, 0·8.

$$\Sigma d = 1\cdot 8, \qquad \therefore \qquad \bar{x} = 12\cdot 45. \qquad \Sigma d^2 = 1\cdot 06.$$

$$s^2 = \frac{1}{n-1}\Sigma(d - \bar{d})^2 = \frac{n\Sigma d^2 - (\Sigma d)^2}{n(n-1)}$$

$$= \frac{4 \times 1\cdot 06 - 3\cdot 24}{4 \times 3}$$

$$= \frac{0\cdot 25}{3}$$

$$t = \frac{12 \cdot 45 - 12 \cdot 0}{[0 \cdot 25/(3 \times 4)]^{\frac{1}{2}}}$$

$$= 0 \cdot 45 \times 4\sqrt{3}$$

$$= 3 \cdot 12$$

From the tables $t_{5\%}(3) = 3 \cdot 18$, $3 \cdot 12 < 3 \cdot 18$ therefore we cannot reject the hypothesis $\mu = 12 \cdot 0$ at the 5 per cent level of significance.

Example 9.14. A random sample of size 12 had a mean $\bar{x} = 14 \cdot 3$ and a variance $s^2 = 2 \cdot 1$. Test at the 5 per cent level of significance that the mean of the population $\mu = 15 \cdot 0$ against the alternative hypothesis $\mu < 15 \cdot 0$.

$$t = \frac{14 \cdot 3 - 15 \cdot 0}{\sqrt{(2 \cdot 1/12)}} = -2\sqrt{0 \cdot 7}$$

$$= -1 \cdot 68$$

Since the tables are two-tailed and we are considering a single-tailed test we must look up the 10 per cent value of t so that we can carry out a 5 per cent single-sided test.

$$t_{10\%}(11) = 1 \cdot 8 \qquad 1 \cdot 68 < 1 \cdot 8$$

therefore we cannot reject the null hypothesis in favour of the alternative.

9.8. TESTS FOR THE DIFFERENCE BETWEEN TWO POPULATION MEANS GIVEN TWO SMALL SAMPLES (POPULATION VARIANCES UNKNOWN)

(*a*) Variances assumed equal ($\sigma_1^2 = \sigma_2^2 = \sigma^2$).

As was the case for large samples we form

$$s^2 = \frac{(n_1 - 1) s_1^2 + (n_2 - 1) s_2^2}{n_1 + n_2 - 2}$$

as an estimate of σ^2.

$$t = \frac{\bar{x}_1 - \bar{x}_2 - (\mu_1 - \mu_2)}{s\sqrt{(1/n_1 + 1/n_2)}}$$

is tested against $t_{\alpha\%}(n_1 + n_2 - 2)$ to test a null hypothesis about $\mu_1 - \mu_2$.

Example 9.15. Two samples of size 6 and 8 have respective means of $\bar{x}_1 = 4 \cdot 25$, $\bar{x}_2 = 4 \cdot 35$ and variances $s_1^2 = 0 \cdot 0025$, $s_2^2 = 0 \cdot 0016$. Test if the samples could come from the same population.

$$s^2 = \frac{5 \times 0.0025 + 7 \times 0.0016}{6 + 8 - 2} = \frac{0.0237}{12}$$

The null hypothesis is $\mu_1 - \mu_2 = 0$.

$$t = \frac{4.25 - 4.35}{\sqrt{[(0.0237/12)(\frac{1}{6} + \frac{1}{8})]}} = -0.1 \sqrt{\left(\frac{288}{0.1659}\right)}$$

$$= -4.189$$

$t_{1\%}(12) = 3.05$, $4.189 > 3.05$ therefore the test is significant at the 1 per cent level and we conclude that the samples come from different populations.

Note: whether or not s_1^2 and s_2^2 are both estimates of σ^2 can be tested using the F test as shown in Chapter 12.

Example 9.16. In *Table 9.6* are two sets of 10 determinations of the percentages of chlorine in each of two batches of polymer. Have the batches a different percentage content of chlorine?

Table 9.6

Batch I	68·59	68·45	69·64	68·64	68·00	67·03
	67·33	67·80	68·04	68·41		
Batch II	65·71	66·65	66·72	67·56	68·27	66·58
	67·08	67·13	67·92	66·21		

Using a false origin of 65 in *Table 9.7* we have

Table 9.7

	I	II
Σd	31·93	19·83
Σd^2	106·8293	44·7817
$\Sigma(d - d)^2$	4·8768	5·4588

$$\bar{x}_1 - \bar{x}_2 = \bar{d}_1 - \bar{d}_2 = 1.210$$

$$s_2 = \frac{4.8768 + 5.4588}{18} = 0.5743$$

$$t = \frac{1.210}{\sqrt{[0.5743 (\frac{1}{10} + \frac{1}{10})]}} = 1.210 \times \sqrt{\left(\frac{5}{0.5743}\right)}$$

$$= 3.56$$

$t_{1\%}$ (18) = 2·89, 3·56 > 2·89 therefore using a 1 per cent level of significance we conclude that the batches have a different percentage content.

(b) Variances are unknown and unequal $\sigma_1^2 \neq \sigma_2^2$.

This gives the celebrated Fisher–Behrens problem which requires special tables to solve it since the statistic

$$t = \frac{\bar{x}_1 - \bar{x}_2 - (\mu_1 - \mu_2)}{\sqrt{(s_1^2/n_1 + s_2^2/n_2)}} \qquad \dots (9.13)$$

no longer follows the t distribution.

An approximate test which is much simpler to use than the full test has been given by Welch and Aspin. It seems to give a satisfactory answer for most practical problems. The test makes use of the t distribution as follows: the statistic given in equation (9.13) has approximately the t distribution with v degrees of freedom where v is given by

$$\frac{1}{v} = \frac{1}{n_1 - 1}\left(\frac{s_1^2/n_1}{s_1^2/n_1 + s_2^2/n_2}\right)^2 + \frac{1}{n_2 - 1}\left(\frac{s_2^2/n_2}{s_1^2/n_1 + s_2^2/n_2}\right)^2$$

Example 9.17. The results of four analyses carried out by two different methods are given in *Table 9.8.*

Table 9.8

Analysis	1	2	3	4
Method I	3·65	3·64	3·65	3·63
Method II	3·68	3·69	3·69	3·68

Test if the methods give significantly different results. In *Table 9.9* using a false origin of 3·6 we have that

Table 9.9

	Method I	Method II
Σd	17×10^{-2}	34×10^{-2}
Σd^2	75×10^{-4}	290×10^{-4}
$\Sigma d^2 - \dfrac{(\Sigma d)^2}{n}$	$2·75 \times 10^{-4}$	1×10^{-4}
$\dfrac{s^2}{n}$	$\dfrac{2·75 \times 10^{-4}}{3 \times 4}$	$\dfrac{1 \times 10^{-4}}{3 \times 4}$

$$t = \frac{8 \cdot 5 - 4 \cdot 25}{\sqrt{(2 \cdot 75/12 + 1/12)}} \times \frac{10^{-2}}{10^{-2}} = 4 \cdot 25 \sqrt{\left(\frac{12}{3 \cdot 75}\right)}$$

$$= 7 \cdot 603$$

$$\frac{1}{v} = \frac{1}{3}\left(\frac{2 \cdot 75/4}{2 \cdot 75/4 + 1/4}\right)^2 + \frac{1}{3}\left(\frac{1/4}{2 \cdot 75/4 + 1/4}\right)^2$$

$$v \simeq 5$$

$t_{1\%}(5) = 4 \cdot 03$, $7 \cdot 603 > 4 \cdot 03$, therefore the methods give significantly different results.

9.9. AN APPROXIMATE METHOD FOR TESTING IF TWO SAMPLES COME FROM POPULATIONS WITH EQUAL MEANS (SAMPLE SIZES SMALL AND EQUAL)

The sample sizes *must* be small and equal,

$$\text{i.e. } n_1 = n_2 = n \leqslant 10$$

The statistic $t^{11} = \dfrac{|\bar{x}_1 - \bar{x}_2|}{R_1 + R_2}$, where R_1 and R_2 are the ranges of the samples, follows the distribution given in *Table 9.10* if $\sigma_1^2 = \sigma_2^2$ and the populations are normally distributed. The test is two-tailed.

Table 9.10. Values of t^{11}

Sample size (n)	Significance level $\alpha = 0 \cdot 05$	$\alpha = 0 \cdot 01$
2	1·714	3·958
3	0·636	1·046
4	0·406	0·618
5	0·306	0·448
6	0·250	0·357
7	0·213	0·300
8	0·186	0·26
9	0·167	0·232
10	0·152	0·210

Consider Example 9.17

$$\bar{x}_1 = 3 \cdot 6425 \quad \bar{x}_2 = \cdot 685$$
$$R_1 = 0 \cdot 02 \quad R_2 = 0 \cdot 01$$
$$t^{11} = \frac{|3 \cdot 6425 - 3 \cdot 685|}{0 \cdot 03} = 1 \cdot 42$$

$t_{5\%}$ (4) = 0·406, 1·42 > 0·406 therefore the results are significantly different. This agrees with the result of Example 9.17.

9.10. TEST FOR PAIRED DATA GIVEN SMALL SAMPLES (POPULATION VARIANCES UNKNOWN)

The normal distribution used in section 9.4 is replaced by the t distribution and we use $(n - 1)$ degrees of freedom. Also $\sigma_1^2 + \sigma_2^2$ is replaced by $s^2 = \frac{\Sigma(d - \bar{d})^2}{(n - 1)}$.

Example 9.18. Use Example 9.7 but suppose that the population variances are unknown.

$$\Sigma d = 1 \cdot 5, \bar{d} = 0 \cdot 3$$
$$\Sigma d^2 = 0 \cdot 16 + 0 \cdot 25 + 0 \cdot 01 + 0 \cdot 01 + 0 \cdot 36 = 0 \cdot 79$$
$$t = \frac{0 \cdot 3}{\sqrt{(0 \cdot 085/5)}} = 0 \cdot 3 \times \sqrt{58 \cdot 82} = 2 \cdot 301$$

$t_{5\%}$ (4) = 2·13 (this is a single-tailed test), 2·301 > 2·13 therefore there is evidence that the compound is effective.

9.11. COMPARISON OF MORE THAN TWO MEANS

Suppose that we have k independent samples each containing n_i ($i = 1, 2, \ldots k$) observations. The mean of the ith sample is $\bar{x}_i$, which is an estimate of the ith population mean μ_i.

In order to carry out any tests of the population means we make the assumption that the population variances are equal. $\sigma_i^2 = \sigma^2$ for all i.

The best estimate of σ^2 is given by

$$s^2 = \frac{\Sigma(x - \bar{x}_1)^2 + \Sigma(x - \bar{x}_2)^2 + \ldots + \Sigma(x - \bar{x}_k)^2}{n_1 + n_2 + \ldots n_k - k}$$

Before we compare any two means we must decide how the two means were chosen. If the two means μ_a and μ_b could have been picked beforehand by other than statistical reasons then we can compare

$$t = \frac{x_a - x_b - (\mu_a - \mu_b)}{s\sqrt{(1/n_a + 1/n_b)}}$$

with $t_{\alpha\%} (n_1 + n_2 + \ldots + n_k - k)$. The result will be applicable only to the two means chosen.

Example 9.19. We have 10 samples one of which is a standard sample. $s = 0.048$ and $\sum\limits_{i=1}^{10} n_i = 50$. The standard sample gives a mean $\bar{x}_{ss} = 17.4$ based on six values. Another sample has a mean $\bar{x} = 17.9$ based on four values. Is this other sample from a population with a different mean to the mean of the standard population?

$$t = \frac{\bar{x} - \bar{x}_{ss}}{s\sqrt{(1/n + 1/n_{ss})}} = \frac{17.9 - 17.4}{0.048\sqrt{(\frac{1}{4} + \frac{1}{6})}} = 5.10$$

$n_1 + n_2 + \ldots + n_k - k = 50 - 10 = 40; t_{5\%}(40) = 2.02, 5.10 > 2.02$ therefore the mean of the other population differs from the mean of the standard population.

If the two means are chosen after the experiment because they give the largest and smallest means, we must remember that if we have k means, there are $^kC_2$ comparisons we could carry out and we will have more chance of obtaining a significant difference.

9.12. CONFIDENCE LIMITS

If we have an estimate of μ and either σ^2 or an estimate of σ^2 then we can find a range of values for μ_0 for which the testing of the null hypothesis $\mu = \mu_0$ against the alternative hypothesis $\mu \neq \mu_0$ will give a non-significant result. Let θ be the estimate of μ (θ can be a single value or a sample mean).

(a) σ^2 Known
To test $\mu = \mu_0$ at the 5 per cent level against $\mu \neq \mu_0$ we form $z = \dfrac{\theta - \mu}{\text{s.d.}\theta}$ ('s.d.θ' is the standard deviation of θ) and test if $z < 1.96$. Thus to obtain a non-significant result at the 5 per cent level we must have

$$-1.96 < \frac{\theta - \mu}{\text{s.d.}\theta} < 1.96$$

i.e. $-1.96 \times \text{s.d.}\theta < \theta - \mu < 1.96 \times \text{s.d.}\theta$

i.e. $\theta - 1.96 \times \text{s.d.}\theta < \mu < \theta + 1.96 \times \text{s.d.}\theta$

we say that $\theta \pm 1.96 \times$ s.d.θ gives a 95 per cent confidence interval for μ.

A 99 per cent confidence interval is given by $\theta \pm 2.58 \times$ s.d.θ.

In general the $(100 - 2\alpha)$ per cent confidence interval is given by $\theta \pm z_\alpha \times$ s.d.θ.

If θ is just a single value x the $(100 - 2\alpha)$ per cent confidence interval is $x \pm z_\alpha \sigma$.

If θ is a mean $\bar{x}$ based on a sample size n we have that the $(100 - 2\alpha)$ per cent confidence interval is $\bar{x} \pm z_\alpha(\sigma/\sqrt{n})$. Diagrammatically we have (*Figure 9.4*). Any normal distribution with a mean between μ_L and μ_U will give a non-significant result at the α per cent level of significance $\bar{x}$ is tested to see if it is a member.

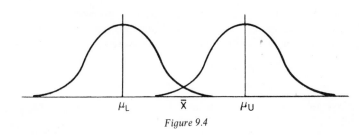

Figure 9.4

Exercises 9e

1. A value of the variable x is found to be 30. It is known to come from a population whose variance is $\sigma^2 = 9$. Find a 95 per cent confidence interval for the population mean μ. (Assume that the population is normal.)

2. A sample of size 25 has a mean of 30 and comes from a population whose variance is 9. What are the 95 per cent confidence limits for μ?

3. The standard deviation of the weight of all males is 20 lb. A random sample of 100 men taken in a town had a mean weight of 162 lb. Find a 98 per cent confidence interval for the mean weight of all males.

(b) σ^2 Unknown and therefore Estimated by s^2, the Sample Variance

For large samples we proceed as in (*a*) but for small samples we must replace the normal distribution by the t distribution. Thus if s^2 is based on v degrees of freedom the $(100 - \alpha)$ per cent confidence interval for μ is $\bar{x} \pm t_\alpha(v) \times \dfrac{s}{\sqrt{n}}$. It is $(100 - \alpha)$ per cent and not $(100 - 2\alpha)$ per cent since the t distribution is given as a double-tailed test in the tables.

Example 9.20. Find a 90 per cent confidence interval for the population mean given the following random sample: 8·41, 8·27, 8·35, 8·29, 8·37, 8·38, 8·49, 8·33, 8·44. Using a false origin 8·00 we have that

$$\bar{x} = 8 + \frac{3·33}{9} = 8·37$$

$$s^2 = \frac{1}{8}\left(1·2715 - \frac{3·33^2}{9}\right) = 0·003925, \, s = 0·0627$$

The 90 per cent confidence interval is given by

$$8·37 \pm \frac{1·86 \times 0·0627}{3}, \text{i.e. } 8·331–8·409.$$

Example 9.21. Find a 95 per cent confidence interval for the difference between the means of the two populations from which the two samples having the following characteristics were selected:

$$\text{Sample I } \bar{x} = 3·4, n = 8, s = 1·4$$
$$\text{Sample II } \bar{x} = 2·6, n = 6, s = 1·2$$

We must first assume a common population variance which we estimate by

$$s^2 = \frac{7 \times 1·96 + 5 \times 1·44}{8 + 6 - 2} = 1·74$$

The 95 per cent confidence interval is given by

$$\bar{x}_1 - \bar{x}_2 \pm t_{5\%}(12)\, s \sqrt{\left(\frac{1}{n_1} + \frac{1}{n_2}\right)}$$

i.e.
$$3·4 - 2·6 \pm 2·18 \sqrt{\left[1·74\left(\frac{1}{6} + \frac{1}{8}\right)\right]}$$

i.e.
$$0·8 \pm 1·55, \text{ i.e. } -0·75–2·35$$

EXERCISES 9

1. For a Student's t distribution with 12 degrees of freedom find the value t_0 of t such that:
    (a) $\Pr(t > t_0) = 0·01$,
    (b) $\Pr(t < t_0) = 0·95$,
    (c) $\Pr(t < t_0) = 0·05$,
    (d) $\Pr(-t_0 < t < t_0) = 0·9$.
2. If a test is two-tailed use the tables to find the values of (a) $t_{5\%}(14)$, (b) $t_{1\%}(9)$, (c) $t_{10\%}(29)$, (d) $t_{2\%}(18)$.

3. If a test is single-tailed use the tables to find the appropriate values of (a) $t_{5\%}$ (19), (b) $t_{0.5\%}$ (23), (c) $t_{1\%}$ (10).

4. Five patients given a sleeping drug took the following times to recover: 1·5 h, 1·7 h, 1·2 h, 1·65 h, 1·55 h; set up (a) a 99 per cent and (b) a 95 per cent confidence interval for the mean recovery time.

5. In an experiment using identical twins pigs, one pig in each of 10 sets was fed foodstuff A while the other twin was fed foodstuff B. The gains in weight in pounds are given in Table 9.11 to the nearest 1 lb.

Table 9.11

A	24	28	31	30	25	27	37	31	26	29
B	19	24	32	28	28	29	30	33	29	27

Test at the 5 per cent level if foodstuff A produces a bigger increase in weight than foodstuff B.

6. Boys entering the army were given an I.Q. test and then 16 were chosen in 8 pairs. Each pair was chosen because they had the same I.Q. and one member of each pair had been in the Boys' Brigade. After 2 weeks the 16 were given an initiative test. Table 9.12 gives the results.

Table 9.12

Pair No.	1	2	3	4	5	6	7	8
B.B.	82	73	66	63	82	61	71	63
Not B.B.	77	73	62	67	64	76	63	68

Is there evidence at the 5 per cent level that the B.B. training has helped in the initiative training?

7. In order to test the yielding properties of a new variety of gooseberry bush compared with the old variety, one of each type is sent to eight regions with varying climatic conditions. They are planted next to each other and the yields are given in Table 9.13.

Table 9.13

Region	1	2	3	4	5	6	7	8
New variety	186	286	106	193	243	192	223	159
Old variety	180	278	109	188	232	190	224	146

Test at the 1 per cent level of significance if the new variety gives a significantly higher yield than the old.

8. A sample of 20 oranges gave 95 per cent confidence limits for the mean weight of the population of an orange grove as 4·63–8·89 oz. Find the sample estimates of the mean and standard deviation of the weight of an orange.

9. A manufacturer using a process to produce plastic plates has found that he obtains 2 per cent rejects. He changes the process and from a sample of 200 plates made, using the new process, he finds that he rejects two plates. Has there been an improvement in the mean number of rejects?

10. A coin is tossed n times and 48 per cent of the time it comes down heads. How large must n be before we can conclude at the 5 per cent level of significance that the coin is biased?

11. The average weekly wage for workers in the car industry is £30, with a standard deviation of £3. A components firm employing 40 men pays them on the average £31.5s.0. per week. Can the firm claim to be paying superior wages to the rest of the industry?

12. A manufacturer of radio valves has found over the years that they have a mean life of 2050 h with a variance of 400 h. A cheaper metal is used to make the filament and a sample of 100 of these new valves give a mean life of 2045 h. Has the cheaper metal caused an inferior valve on average?

13. Two samples of chain links taken from different manufacturers, using the same method of manufacture, when tested for breaking strength gave results as shewn in *Table 9.14*:

Table 9.14

	Sample I	Sample II
No. in sample	50	70
Sum of breaking strengths	75 ton	140 ton
Sum of squares of breaking strengths	150	300

Do these results indicate at the 1 per cent level that there is a difference in the strengths of the chains supplied by the two manufacturers?

14. A die is rolled 60 times and 25 times the die shows one or two. Is this evidence that the die is biased in favour of ones and twos?

15. A sample of size 14 has a mean of 142·85 and a variance of 47·27. Find the 95 per cent confidence limits for the mean of the population.

16. Two soap powders are tested to see which washes whiter, six articles being washed with each powder. The results are the measurements on a machine which measures the whiteness of the articles washed. See *Table 9.15.*

Table 9.15

Brand X	38	36	40	38	37	39
Zad	21	23	22	21	22	23

Is there evidence to support the theory that Zad is an inferior soap powder?

17. Two batches of steel were sampled for carbon content and the results are given in *Table 9.16.*

Table 9.16

Batch I	4·6	5·3	4·8	5·1	4·7
Batch II	4·8	5·0	5·2	5·3	4·9

Test at the 1 per cent level of significance whether
(a) the batches contain different amounts of carbon,
(b) Batch II contains more carbon than Batch I.

18. Two batches of chickens A and B (chosen because of equal egg production) were subjected to different types of environment. A had 'pop' music played for 5 min every half hour while B had classical music played for 5 min every half hour. The number of eggs produced in a given week is to be found in *Table 9.17.*

Table 9.17

Batch A	50	60	47	55	59	49	54	54	56	51
Batch B	45	55	49	50	49	48	48	50	49	

Test if 'pop' music is better than classical music in increasing egg production.

19. A random sample of 500 Liverpool University students was selected, and each student asked whether his or her home was inside the City boundaries. Three hundred and fifty affirmative replies were obtained. Assuming that the total number of students is 3300, estimate the number of those who live within the City boundaries, and give a confidence interval for your estimate.

Explain the method you would adopt if asked to carry out this inquiry.   (Liv. U.)

20. Female shrimps were collected from beaches and rock-pools and the number of eggs carried by each shrimp was counted. In the size group 9–10 mm 20 shrimps were collected from the beach and 10 from rock-pools, the individual numbers of eggs is given in *Table 9.18*.

Table 9.18

Locality	Numbers of eggs									
Beach	15,	14,	14,	14,	13,	13,	16,	12,	13,	18
	20,	14,	10,	15,	13,	19,	14,	17,	18,	10
Rock-pool	24,	17,	12,	15,	23,	17,	19,	18,	22,	28

Test whether there is any significant difference between the average numbers of eggs carried in the two localities.   (Liv. U.)

21. Describe what is meant by a null hypothesis.

In a tasting trial a subject is presented with four different samples of real coffee and one of instant coffee and asked if he can identify the instant coffee (the order of presentation of the specimens being random). Of 10 people subjected to the trial, four identify the instant coffee. Is there any evidence at the 5 per cent level of significance, that people are able to discriminate between real coffee and instant coffee?

Investigate also the case of 30 people out of 100 correctly identifying the instant coffee.   (Liv. U.)

22. A firm which manufactures lead-covered submarine cable suspected that the lead was being put on more thickly by its night workers than by its day workers. To keep down the cost the lead has to be as thin as possible but nowhere less than 0·2 in thick if it is to withstand the action of the sea water and general wear and tear. Part of the investigation carried out is summarized in *Table 9.19*.

Table 9.19

	Day work	Night work
No. of places at which the thickness of the lead cover was measured	100	100
Mean thickness (in)	0·292	0·298
Standard deviation (in)	0·021	0·019

Determine whether the difference between the means is significant or not and make comments on the implications of the given figures and your result. (J.M.B.)

23. Large samples of male and female plants of dog's mercury were collected at each of thirteen sites in Derbyshire. The mean numbers of leaf pairs in these samples is given in *Table 9.20*.

Table 9.20

Site	1	2	3	4	5	6	7	8	9	10	11	12	13
Male	6·0	6·6	8·0	7·0	6·4	6·9	6·1	6·9	6·6	8·2	7·9	7·0	7·5
Female	5·8	7·3	6·6	6·7	6·3	6·2	6·1	7·3	6·0	6·7	8·2	6·0	6·5

Find the difference between the mean numbers of leaf pairs of male and female plants from site to site, and calculate the mean and standard deviation of these differences.

State whether or not these data can be taken to establish a real difference between the average numbers of leaf pairs for male and female plants. Give reasons for your statement, making any calculations you think necessary and stating any assumptions you make. (J.M.B.)

24. A sample of n_1 observations has mean m_1 and standard deviation s_1, while an independent sample of n_2 observations has mean m_2 and standard deviation s_2. Show that if n_1 and n_2 are both large, the standard error of $(m_1 - m_2)$ is approximately given by

$$\left(\frac{s_1^2}{n_1} + \frac{s_2^2}{n_2} \right)^{\frac{1}{2}}$$

The annual salaries of a random sample of 1000 individuals of a particular age and profession has mean £1020 and standard deviation £50, while a random sample of 500 individuals of the same age but belonging to a different profession has a mean annual salary of £980 with standard deviation £75. Determine whether there is a significant difference between the average salaries in the two professions at the given age. (J.M.B.)

25. (a) Explain briefly what is meant by the statement that 'a sample mean differs significantly from some assumed value'.

(b) Discuss the merits of using the sample median, sample mode or sample mean to estimate a population mean, and also the rôle of sampling distributions in deciding which statistic should be used.
(J.M.B.)

10

QUALITY CONTROL

10.1. INTRODUCTION

WHEN products are being manufactured on a production line basis it is important to keep a check on their quality. If a change in quality occurs the sooner the production line is stopped the smaller the amount of unacceptable product that is manufactured.

One of the simplest ways of checking is to construct *control charts*. These are graphs on which some measure of the quality of a product, estimated by taking samples, is recorded as manufacture is proceeding. The charts consist of a central line and two pairs of limit lines spaced above and below the central line.

—————————————————————Outer control limit

—————————————————————Inner control limit

—————————————————————Central line

—————————————————————Inner control limit

—————————————————————Outer control limit

Figure 10.1

Control charts enable one to test whether the samples are random samples from the same population or whether changes in the population have occurred.

In general, control charts are drawn up for sample means (or proportions) and sample ranges, since either the population mean or the variability about the mean could change.

There is bound to be some variation due to chance causes which are inherent in the production system and which cannot be eliminated. If these are the only causes of variation then the production is said to be *under control*.

164

10.2. CONTROL CHARTS FOR SAMPLE MEANS

For samples of size n the means are approximately normally distributed mean μ, variance σ^2/n where μ and σ^2 are the population mean and variance. Thus 95 per cent of the sample means will lie in the interval

$$\mu \pm 1.96\frac{\sigma}{\sqrt{n}} \text{ (the inner limits)}$$

while 99·8 per cent will lie in the interval

$$\mu \pm 3.09\frac{\sigma}{\sqrt{n}} \text{ (the outer limits)}$$

Thus we can expect 1 sample in 20 to give an answer outside the inner limits and 1 in 500 outside the outer limits.

If μ and σ^2 are unknown they have to be estimated. An initial batch of samples is taken and used to find estimates of μ and σ^2 and when the process has had a chance to settle down another batch of samples is taken to find new estimates.

Using the initial batch of samples the estimate for μ is given by $\bar{x}$, the mean of all the members of the samples.

Example 10.1. The results of twelve samples are given in *Table 10.1.*

Table 10.1

Sample No.	1	2	3	4	5	6	7	8	9	10	11	12
	4	0	2	9	2	4	2	1	9	4	2	8
	6	4	5	1	8	4	0	6	7	3	4	1
	5	1	3	0	7	5	3	8	8	1	5	7
	2	8	4	2	6	5	1	3	7	2	9	2
Sum	17	13	14	12	23	18	6	18	31	10	20	18

Range 4 8 3 9 6 1 3 7 2 3 7 7

Total sum = 200, $\therefore \bar{x} = \dfrac{200}{48} = 4.166^{\cdot}$

The estimate of σ can be obtained in one of three ways:

(a) The most accurate estimate is given by the square root of

the *variance within samples* = $\sqrt{\left[\dfrac{\sum_i \sum_j (x_{i,j} - \bar{x}_i)^2}{k(n-1)}\right]}$

where $\bar{x}_i$ is the mean of the ith sample; there are k samples and n members in each sample. Using the identity

$$\sum_i \sum_j (x_{i,j} - \bar{x}_i)^2 = \sum_i \sum_j x_{i,j}^2 - \sum_i \frac{(\sum_j x_{i,j})^2}{n}$$

the estimate of σ^2 in Example 11.1 is

$$\frac{1}{12 \times 3} \left(1178 - \frac{3796}{4}\right) = \frac{229}{36} = 6\cdot631$$

From this the estimate of σ is $2\cdot52$.

(b) σ can also be estimated by the average of the within samples standard deviation, then we must evaluate

$$s = \sqrt{\left[\frac{\Sigma(x - \bar{x})^2}{(n-1)}\right]} = \sqrt{\left[\frac{n\Sigma x^2 - (\Sigma x)^2}{n(n-1)}\right]}$$

for all the samples. This is done most easily by means of *Table 10.2*.

Table 10.2

Sample	Σx (1)	Σx^2 (2)	$(\Sigma x)^2$ (3)	$4\Sigma x^2$ (4)	$4\Sigma x^2 - (\Sigma x)^2$ (4) − (3)	$\dfrac{(4)-(3)}{12}$	$\sqrt{\left[\dfrac{(4)-(3)}{12}\right]}$
1	17	81	289	324	35	2·9167	1·7079
2	13	81	169	324	155	12·9167	3·5939
3	14	54	196	216	20	1·6667	1·2907
4	12	86	144	344	200	16·6667	4·0832
5	23	153	529	612	83	6·9167	2·63
6	18	82	324	328	4	0·3333	0·5773
7	6	14	36	56	20	1·6667	1·2907
8	18	110	324	440	116	9·6667	3·1092
9	31	243	961	972	11	0·9167	0·9574
10	10	30	100	120	20	1·6667	1·2907
11	20	126	400	504	104	8·6667	2·944
12	18	118	324	472	148	12·3333	3·5116
						Total	26·9866

$$\bar{s} = \frac{26 \cdot 9866}{12} = 2 \cdot 249$$

(c) If we assume the population is normally distributed we can estimate σ from the mean range $\bar{w}$, provided the sample size is less than or equal to 12 and the number of samples is greater than 10. The estimate of σ is given by $a_n \bar{w}$ where a_n is given in *Table 10.3* for values of n from 2 to 9.

Table 10.3

n	2	3	4	5	6	7	8	9
a_n	0·8862	0·5908	0·4857	0·4299	0·3946	0·3698	0·3512	0·3367

If less than 10 samples are available to find the first estimate of σ, a fresh estimate should be calculated when more samples are available.

For Example 10.1 $\qquad \bar{w} = \dfrac{60}{12} = 5, a_4 = 0 \cdot 4857$

$\therefore$ the estimate of σ is $5 \times 0 \cdot 4857 = 2 \cdot 43$.

The third method is the most usual one to be used as it is the simplest. It is the one used in all the problems considered in this chapter.

Using this method for Example 10.1, the inner control limits are given by

$$\bar{x} \pm 1 \cdot 96 \frac{a_n}{\sqrt{n}} \bar{w} = \bar{x} \pm A_{0 \cdot 025} \bar{w}$$

$$= \bar{x} \pm A_{0 \cdot 025} \bar{w} \text{ where } A_{0 \cdot 025} = \frac{1 \cdot 96 a_n}{\sqrt{n}}$$

and the outer control limits are given by

$$\bar{x} \pm 3 \cdot 09 \frac{a_n}{\sqrt{n}} \bar{w}$$

$$= \bar{x} \pm A_{0 \cdot 001} \bar{w} \text{ where } A_{0 \cdot 001} = \frac{3 \cdot 09 a_n}{\sqrt{n}}$$

Values of $A_{0 \cdot 025}$ and $A_{0 \cdot 001}$ are given for different values of n in *Table 10.4*.

Thus, for Example 10.1 the inner control limits are

$$4 \cdot 16 \pm 0 \cdot 476 \times 5, \text{ i.e. from } 1 \cdot 78 \text{ to } 6 \cdot 54$$

and the outer control limits are

$$4 \cdot 16 \pm 0 \cdot 75 \times 5, \text{ i.e. from } 0 \cdot 41 \text{ to } 7 \cdot 91$$

The means for each sample are plotted as soon as they have been worked out. If a value comes outside the inner limits but inside the outer limits it is taken as a warning and a further sample is taken as soon as possible to check if this suspect result is the allowable 1 in 20.

Table 10.4

n	2	3	4	5	6	7	8	9	10	11	12
$A_{0 \cdot 025}$	1·229	0·608	0·476	0·377	0·316	0·274	0·244	0·220	0·202	0·186	0·174
$A_{0 \cdot 001}$	1·937	1·054	0·750	0·594	0·498	0·432	0·384	0·347	0·317	0·294	0·274

If a point comes outside the outer limits (the chances of this are 1 in 500) the production is stopped and the production line is checked for faults. From the above procedures it is easily seen why the inner limits are called 'warning limits' and the outer limits are called 'action limits'.

For Example 10.1 the sample means are 4·25, 3·25, 3·5, 3·0, 5·75, 4·50, 1·5, 4·5, 7·75, 2·5, 5·0 and 4·5 and these are shown plotted in *Figure 10.2*.

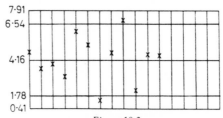

Figure 10.2

No points are outside the action limits but two points are outside the warning limits. However, they are on opposite sides of the mean, and further samples indicate that they are just chance results.

Even if the means are between the control limits the variation could be large and a further chart is required to keep a check on the variation. The most suitable and easiest chart to form is described in section 10.3.

10.3. CONTROL CHARTS FOR RANGES

The inner control limits for ranges are given by $F_{0 \cdot 025}\sigma - F_{0 \cdot 975}\sigma$ and the outer control limits are $D_{0\cdot001}\sigma$–$D_{0\cdot999}\sigma$. They correspond to 95 per cent and 99·8 per cent confidence limits for the range. If σ is estimated, using $\bar{w}$, the inner limits are $F'_{0\cdot025}\bar{w}$–$F'_{0\cdot975}\bar{w}$ and the outer limits are $D'_{0\cdot001}\bar{w}$–$D'_{0\cdot999}\bar{w}$. Values of F, D, F' and D' are given in *Table 10.5* for different values of n.

Table 10.5

	For use with σ				For use with $\bar{w}$			
n	$D_{0\cdot001}$	$F_{0\cdot025}$	$F_{0\cdot975}$	$D_{0\cdot999}$	$D'_{0\cdot001}$	$F'_{0\cdot025}$	$F'_{0\cdot975}$	$D'_{0\cdot999}$
2	0·00	0·04	3·17	4·65	0·00	0·04	2·81	4·12
3	0·06	0·30	3·68	5·06	0·04	0·18	2·17	2·99
4	0·20	0·59	3·98	5·31	0·10	0·29	1·93	2·58
5	0·37	0·85	4·20	5·48	0·16	0·37	1·81	2·36
6	0·54	1·06	4·36	5·62	0·21	0·42	1·72	2·22
7	0·69	1·25	4·49	5·73	0·26	0·46	1·66	2·12
8	0·83	1·41	4·61	5·82	0·29	0·50	1·62	2·04
9	0·96	1·55	4·70	5·90	0·32	0·52	1·58	1·99
10	1·08	1·67	4·79	5·97	0·35	0·54	1·56	1·94
11	1·20	1·78	4·86	6·04	0·38	0·56	1·53	1·90
12	1·30	1·88	4·92	6·09	0·40	0·58	1·51	1·87

As for control charts for means, the limits are warning and action limits. For Example 10.1 the inner limits are

$$0\cdot29 \times 5\text{–}1\cdot93 \times 5$$

i.e. $\qquad\qquad\qquad 1\cdot45\text{–}9\cdot65$

The outer limits are

$$0\cdot1 \times 5\text{–}2\cdot58 \times 5$$

i.e. $\qquad\qquad\qquad 0\cdot5\text{–}12\cdot9$

The ranges for each sample are plotted in *Figure 10.3*. No points are outside the action limits and only one point is outside the warning limits and these limits are only there to check that no bias has been introduced into the sampling so that the process is 'under control'.

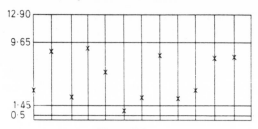

Figure 10.3

Exercises 10a

1. *Table 10.6* gives the deviations in 0·001 in from 1·500 in of the dimensions of an article. Twenty samples of 5 were taken. Find control limits for the mean and range. Plot the means and ranges on the chart and discuss the control of the process.

Table 10.6

2	3	1	1	9	3	6	5	8	5	0	6	2	5	4	2	9	6	9	9
4	3	3	2	2	7	5	8	7	4	4	5	4	5	3	2	8	3	8	9
5	7	5	3	7	6	9	2	3	3	2	8	6	6	7	3	7	5	3	6
6	8	6	4	6	4	8	6	6	5	9	2	8	5	8	5	1	8	7	8
8	4	5	5	6	5	4	4	6	1	7	7	3	5	2	6	2	1	7	9

10.4. CONTROL CHARTS FOR FRACTION DEFECTIVE

The quality of a product cannot always be measured on a continuous scale. It may be that all that can be measured is whether or not the product possesses a certain characteristic. If the product does not possess this characteristic it is called defective. If the probability of a given item being defective is p then we would expect the fraction x/n, where x is the number of defectives in a sample size n, to vary as a binomial distribution with mean p and variance $p(1 - p)/n$.

Note: x follows the binomial distribution with $\mathscr{E}(x) = np$ and $\mathrm{var}(x) = np(1 - p)$.

$$\therefore \qquad \mathscr{E}\left(\frac{x}{n}\right) = \frac{\mathscr{E}(x)}{n} = \frac{np}{n} = p.$$

and

$$\mathrm{var}\left(\frac{x}{n}\right) = \frac{\mathrm{var}(x)}{n^2} = \frac{np(1 - p)}{n^2} = \frac{p(1 - p)}{n}$$

Control charts can be set up using the binomial distribution, but the working is simplified if we use either the Poisson or the normal distribution to approximate the binomial distribution.

If the samples are of unequal size we can use the average sample size n to compute average limits. However, if the variation in sample size is large individual limits should be constructed for a range of sample sizes.

If p is not specified we must estimate it using

$$\hat{p} = \frac{\text{total number of defectives in first few samples}}{\text{total number in the first few samples}}$$

$\hat{p}$ should be re-estimated as more samples are taken and the process settles down.

If the normal approximation is applicable the inner and outer control limits are given by

$$p \pm 1.96 \sqrt{\left[\frac{p(1-p)}{n}\right]}, \quad \text{and} \quad p \pm 3.09 \sqrt{\left[\frac{p(1-p)}{n}\right]}$$

Example 10.2. The samples given in *Table 10.7* had a number of defectives as shown. Find control limits for the fraction defective.

Table 10.7

Sample	1	2	3	4	5	6	7	8	9	10
No. in the sample	120	135	100	110	94	150	106	143	155	124
No. of defectives	10	13	8	12	9	16	12	14	14	6
p	0·0833	0·0962	0·0800	0·1090	0·0957	0·1066	0·1132	0·0979	0·0903	0·0483

Sample	11	12	13	14	15	16	17	18	19	20
No. in the sample	108	126	145	148	118	115	110	121	137	115
No. of defectives	14	13	11	12	15	14	8	10	13	11
p	0·1296	0·1031	0·0758	0·0810	0·1271	0·1217	0·0727	0·0826	0·0949	0·0956

$$\bar{n} = \frac{\Sigma n}{20} = \frac{2480}{20} = 124 \text{ average sample size}$$

$$\bar{d} = \frac{\Sigma d}{20} = \frac{235}{20} = 11\cdot75 \text{ average number of defectives in a sample}$$

$$\hat{p} = \frac{235}{2480} = 0\cdot0948$$

Thus the average control limits are given by

$$0\cdot0948 \pm 1\cdot96 \sqrt{\left(\frac{0\cdot0948 \times 0\cdot9052}{124}\right)} = 0\cdot0948 \pm 1\cdot96 \times 0\cdot0263$$

i.e. $0\cdot0432\text{--}0\cdot1464.$

The average outer control limits are given by

$$0\cdot0948 \pm 3\cdot09 \times 0\cdot0263$$

i.e. $0\cdot0135\text{--}0\cdot1761$

These limits will be smaller for samples larger than average and larger for samples smaller than average. The largest value of n is 155 and the corresponding limits are

$$0\cdot0948 \pm 1\cdot96 \sqrt{\left(\frac{0\cdot0948 \times 0\cdot9052}{155}\right)}$$

and $$0\cdot0948 \pm 3\cdot09 \sqrt{\left(\frac{0\cdot0948 \times 0\cdot9052}{155}\right)}$$

$$= 0\cdot0948 \pm 1\cdot96 \times 0\cdot02353 \text{ and } 0\cdot0948 \pm 3\cdot09 \times 0\cdot02353$$

i.e. $0\cdot0487\text{--}0\cdot1409 \text{ and } 0\cdot0221\text{--}0\cdot1675.$

The smallest value of n is 94 and the corresponding limits are

$$0\cdot0948 \pm 1\cdot96 \sqrt{\left(\frac{0\cdot0948 \times 0\cdot9052}{94}\right)}$$

and $$0\cdot0948 \pm 3\cdot09 \sqrt{\left(\frac{0\cdot0948 \times 0\cdot9052}{94}\right)}$$

i.e. $0\cdot0356\text{--}0\cdot154 \text{ and } 0\cdot0014\text{--}0\cdot1882$

These two calculations give the upper and lower limits for the values of the interval. It may be necessary to calculate limits for n between the largest and the smallest if the proportion of defectives in the sample is close to the average limit lines.

Note: a negative value for one of the limits means that, for samples of the size used, zero defectives can occur in a sample without cause to doubt that the sample is a random one (this is the reason why the lower limits are included).

If the mean number of defectives per sample is less than six then the Poisson approximation to the binomial distribution should be used. It will be assumed that the samples are of the same size and so control charts for the number of defectives per sample can be drawn up.

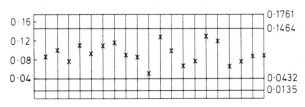

Figure 10.4

Example 10.3. The number of defectives in 20 samples of size 20 are: 2, 1, 0, 2, 1, 0, 0, 1, 1, 2, 0, 1, 0, 1, 0, 3, 0, 2, 1, 0. Set up a control chart for the number of defectives in a sample.

Total number of defectives $= 18$

$\therefore$ the mean number of defectives per sample $= 0.9$

Using the Poisson distribution

		Cumulative probability
$\Pr(0 \text{ def.}) = e^{0.9}$	$= 0.4066$	0.4066
$\Pr(1 \text{ def.}) = 0.9 \Pr(0)$	0.3659	0.7725
$\Pr(2 \text{ def.}) = \dfrac{0.9}{2} \Pr(1) = 0.1647$		0.9372
$\Pr(3 \text{ def.}) = \dfrac{0.9}{3} \Pr(2) = 0.0494$		0.9866
$\Pr(4 \text{ def.}) = \dfrac{0.9}{4} \Pr(3) = 0.0111$		0.9977
$\Pr(5 \text{ def.}) = \dfrac{0.9}{5} \Pr(4) = 0.0020$		0.9997

We can set up control limits for c (number of defectives in a sample) such that the probability of there being c or more defectives is (*a*) less than 0.05 and (*b*) less than 0.01. Since the probability of getting zero defectives is so large there are no lower limits.

Pr(3 or more def.) = $1 - 0.9372 = 0.0618 > 0.05$
Pr(4 or more def.) = $1 - 0.9866 = 0.0134 < 0.05$
Pr(5 or more def.) = $1 - 0.9977 = 0.0023 < 0.01$

Thus the inner limit is 4 and the outer limit is 5.
Referring to *Figure 10.5* we see that the process is under control.

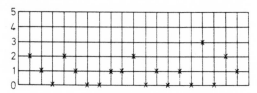

Figure 10.5

Exercises 10b

1. In a process making electric light bulbs a sample of size 150 is taken every half hour and tested. The number of defective bulbs found in the first 10 samples is given in *Table 10.8*. Set up a control chart for the fraction of defectives in a sample.

Table 10.8

Sample	1	2	3	4	5	6	7	8	9	10
No. def.	7	3	8	10	5	8	11	6	8	7

2. Samples, size 16, are taken of valves being manufactured on a continuous process. The number of defectives in the first 15 samples is given in *Table 10.9*. Set up control limits for the number of defectives in a sample.

Table 10.9

Sample	1	2	3	4	5	6	7	8	9	10	11	12	13	14	15
No. def.	0	1	2	1	0	0	2	0	0	0	1	0	0	2	0

Is the process under control?

10.5. ALLOWABLE WIDTH OF CONTROL LIMITS WHEN TOLERANCE LIMITS ARE SPECIFIED

The outer control limits for the means of samples of size n are $\bar{x} \pm 3.09(\sigma/\sqrt{n})$. When $n = 1$ this becomes $\bar{x} \pm 3.09\sigma$. Thus we can see that even if the means of samples of size n are inside the outer control limits we can have individual values outside these limits. The distance between the outer control limits for individual values is 6.18σ and so the upper and lower tolerance limits T_u and T_l must be at least 6.18σ apart to enable us to have the process under control such that there is only 1 chance in 500 of getting a value outside the control limits.

If $T_u - T_l > 6.18\sigma$ $(= 2\sqrt{n}A_{0.001}\bar{w})$ we can find upper and lower confidence limits for means by subtracting $3.09\sigma - 3.09\sigma/\sqrt{n}$ from T_u and adding it to T_l. This is to allow individual values outside the control limits, but not outside the tolerance limits when the means are under control.

If σ is estimated using $\bar{w}$ we have that

$$3.09\sigma - 3.09\frac{\sigma}{\sqrt{n}} = (\sqrt{n} - 1)3.09\frac{a_n}{\sqrt{n}}\bar{w} = (\sqrt{n} - 1)A_{0.001}\bar{w}$$

Example 10.4. Consider Example 10.1, $n = 4$, $\bar{w} = 5$.

$$(\sqrt{n} - 1)A_{0.001}\bar{w} = (2 - 1)0.75 \times 5 = 3.75.$$

Suppose that the tolerance limits are from 0 to 16,

i.e. $\qquad\qquad\qquad T_u = 16$, and $T_l = 0$.

$$6.18\sigma = 6.18\sqrt{n}\frac{a_n}{\sqrt{n}}\bar{w} = 2\sqrt{n}\,3.09\frac{a_n}{\sqrt{n}}\bar{w}$$

$$= 2\sqrt{n}\,A_{0.001}\bar{w} = 2 \times 2 \times 0.75 \times 5 = 15$$

The tolerance limits are sufficiently far apart to allow the process to be under control. We shall now consider if, in fact, this is the case with the values given.

Using the tolerances, the outer control limits for the sample means are $16 - 3.75$ and $0.0 + 3.75$

i.e. $\qquad\qquad\qquad\qquad$ 3.75 to 12.25

Some of the means of the given sample fall outside this range so the process is not in control to the given tolerances.

Note: the inner control limits are at a distance

$$(3.09 - 1.96)\frac{\sigma}{\sqrt{n}} = \left(3.09\frac{a_n}{\sqrt{n}} - 1.96\frac{a_n}{\sqrt{n}}\right)\bar{w}$$

$$= (A_{0.001} - A_{0.025})\bar{w}$$

inside the outer control limits.

Exercises 10c

1. If tolerances of 1·490–1·515 are placed on the articles given in Example 10.1, find the outer and inner control limits for the sample means.

EXERCISES 10

1. In the manufacture of a certain chemical product three weekly tests are made of the percentage acid content. Tests for 10 weeks are available (*Table 10.10*). Set up a control chart and show that the production is stable, also that a guarantee of not more than 6·9 per cent of acid in the product would be reasonable.

Table 10.10

Week	1	2	3	4	5	6	7	8	9	10
	5·74	6·21	5·28	5·51	5·19	5·91	5·27	5·40	6·00	5·07
	5·54	5·18	5·68	5·81	6·21	5·56	6·22	5·42	5·41	5·60
	5·97	5·50	5·00	5·75	4·71	4·79	5·85	5·68	5·75	5·20

2. In manufacture, golf balls are tested for bounce on a machine which measures the rebound after they are dropped from 2 ft. A sample of seven golf balls is tested every day. The following are the results of 20 days testing. Use the first 10 samples to set up control charts and plot all samples to see if the process is under control. In *Table 10.11*

Table 10.11

Day No.	1	2	3	4	5	6	7	8	9	10
	10	9	3	1	9	10	4	7	5	3
	6	7	6	0	8	9	7	9	2	5
	4	4	3	1	7	8	3	6	1	1
	8	4	6	2	6	1	8	8	0	6
	2	1	4	3	4	3	6	9	3	8
	3	1	2	4	6	10	2	9	6	2
	5	2	1	3	2	2	1	8	2	1

Day No.	11	12	13	14	15	16	17	18	19	20
	4	2	1	1	3	9	7	9	5	4
	9	6	5	6	6	4	5	6	3	6
	7	8	5	7	9	0	4	2	6	1
	7	5	0	2	8	0	1	6	6	2
	7	9	1	1	6	6	9	1	1	9
	9	4	4	9	8	1	8	6	4	1
	2	1	8	2	6	4	3	9	4	8

the seven values listed under each day give the deviations from 18 in (in 0·1 in) of the rebounds of the seven balls.

3. For a controlled operation on a lathe the dimension of a small part was specified as 1·29 ± 0·05. Samples of five were taken and for 20 samples the mean was 1·288 with a mean range of 0·03. Find the outer and inner control limits for the mean and range of the samples and the allowable width of control limits for the population mean.

4. In 100 samples each of 30 articles the frequency of occurrence of defectives is given in *Table 10.12*. Set up a control chart for the number of defectives. Is the process under control?

Table 10.12

No. defective	0	1	2	3	4
Frequency	31	47	10	5	1

5. Twenty samples of 30 from every 200 articles produced by a machine show the following number of defectives: 1, 3, 4, 5, 0, 4, 3, 1, 2, 6, 1, 4, 0, 4, 3, 5, 3, 2, 1, 0. Calculate the control limits for defectives. Find the percentage of articles that must be sampled to guarantee a final process average of not more than 5 per cent defectives.

6. A sample scheme accepts lots only when there are no defectives in a sample of 30. What proportion of lots containing 5 per cent defective will be accepted?

7. A machine is producing articles to a nominal dimension of 20 in with a tolerance of ±0·020 in. Twenty-five samples of five are taken and the grand mean is found to be 20·005 in and the average range 0·012. Find the outer control limits for mean and range. Assuming that the process is under control, show that the machine will not meet the tolerances unless the mean is lowered by approximately 0·001.

8. A machine is producing articles of nominal weight 25 lb at the rate of 200/h. Samples of eight taken hourly show a mean weight of 25·034 lb and a mean range of 0·061 lb. The process is under control. Show that the machine is working to tolerance limits of ±0·10 lb and that by adjustment it may be possible to work to tolerances of ±0·066 lb.

11

CHI-SQUARED DISTRIBUTION

11.1. INTRODUCTION

WHEN carrying out tests of significance using small samples (population variance unknown) we used the t test. The statistic used was defined as

$$t = \frac{\bar{x} - \mu}{s/\sqrt{n}} = \frac{\bar{x} - \mu}{\sigma/\sqrt{n}} \Big/ \sqrt{\frac{s^2}{\sigma^2}}$$

$$= \frac{\bar{x} - \mu}{\sigma/\sqrt{n}} \Big/ \sqrt{\frac{\chi^2}{v}}$$

where χ^2 follows the chi-squared distribution and v is the number of degrees of freedom associated with s^2. The chi-squared distribution will be considered in this chapter. It can be used as a test of variance, a test of goodness of fit and also to set up a confidence interval for the population variance.

11.2. DEFINITION

If $(z_i) = \left(\dfrac{x_i - \mu}{\sigma}\right)$ $i = 1, 2, \ldots n$ are a sample from the standardized

normal distribution then the sum of squares $\chi^2 = \displaystyle\sum_{i=1}^{n} z_i^2 = \dfrac{vs^2}{\sigma^2}$ has the

probability element $f(\chi^2) \, d\chi^2 = \dfrac{1}{2^{v/2} \, \Gamma(v/2)} (\chi^2)^{v/2-1} e^{-\chi^2/2} \, d(\chi^2) \, (0 \leqslant \chi^2$

$< \infty)$ where v is the number of degrees of freedom associated with s^2 and $\Gamma(v/2)$ is the gamma function.

The probability element is sometimes written:

$$f(\chi^2) \, d\chi^2 = \frac{1}{\Gamma(v/2)} \left(\frac{\chi^2}{2}\right)^{v/2-1} e^{-\chi^2/2} \, d\left(\frac{\chi^2}{2}\right)$$

The distribution so defined is called the chi-squared distribution and has the following properties:

(a) $\int_0^\infty f(\chi^2) \, d\chi^2 = 1$. This shows that it is a probability distribution.
(b) Its mean is v, i.e. $\mathscr{E}(\chi^2) = v$.
(c) Its variance is $2v$, i.e. var $(\chi^2) = \mathscr{E}(\chi^2 - v)^2 = 2v$.

(d) The maximum value of $f(\chi^2)$ occurs when $\chi^2 = v - 2$ for $v \geqslant 2$.

(e) v is the number of independent variables which are used to calculate χ^2.

For example suppose that there are variates $x_1, x_2, \ldots x_n$ but if we have to estimate μ using $\bar{x}$ then we have only $n - 1$ independent variates. If a further m parameters have to be estimated using the original variates $x_1, x_2, \ldots x_n$, then we shall have only $n - m - 1$ independent variables remaining. The number of independent variables is called the number of degrees of freedom.

(f) Figure 11.1 gives a comparison of the shape of the distribution for various values of v.

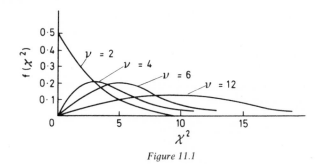

Figure 11.1

(g) Percentage points χ_p^2 are given in the tables for various values of v. χ_p^2 and p are defined in one of two ways; either

$$(a)\ \frac{p}{100} = \int_{\chi_p^2}^{\infty} f(\chi^2)\,d\chi^2 \quad \text{(see Figure 11.2)}$$

$$\text{or } (b)\ \frac{p}{100} = \int_{0}^{\chi_p^2} f(\chi^2)\,d\chi^2 \quad \text{(see Figure 11.3)}$$

In case (a) $p/100$ is the probability that a variable χ^2 is greater than χ_p^2 while in case (b) it is the probability that χ^2 is less than χ_p^2. χ_p^2 is called the critical value of χ^2 for a given v and p.

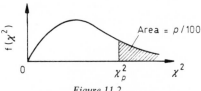

Figure 11.2

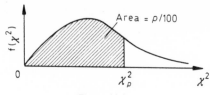

Figure 11.3

In general, tables are only given for $v = 1$ (1) 30 and 30 (10) 100, linear interpolation being adequate in the second range.

For values of $v > 100$ we make use of the fact that $\sqrt{(2\chi^2)}$ tends to the normal distribution as $v \to \infty$. We use $\sqrt{(2\chi^2)}$ and not χ^2 as the convergence is much more rapid. Thus for $v > 100$, $\sqrt{(2\chi^2)}$ is approximately normally distributed with mean $\sqrt{(2v - 1)}$ and unit variance.

A comparison of the respective values of $(\chi^2 - v)/\sqrt{(2v)}$ (the standardized variable for the χ^2 distribution) and $\sqrt{(2\chi^2)} - \sqrt{(2v - 1)}$ (the standardized variable for $\sqrt{(2\chi^2)}$ are shown in *Table 11.1* for the 5 per cent level of p [using definition (*a*)].

Table 11.1

v	$\chi^2_{5\%}$	$\dfrac{\chi^2 - v}{\sqrt{(2v)}}$	$\sqrt{(2\chi^2)} - \sqrt{(2v - 1)}$
1	3·84	2·01	1·77
25	37·7	1·80	1·69
100	124·3	1·72	1·66

From the normal distribution tables, if $\phi(z) = 0.95$, $z = 1.645$. This compares favourably with the value 1·66 obtained for $v = 100$ using $\sqrt{(2\chi^2)} - \sqrt{(2v - 1)}$.

Thus to calculate $\chi^2_{5\%}$ ($v = v_0 > 100$) we use $1.645 = \sqrt{(2\chi^2)} - \sqrt{(2v_0 - 1)}$.

Note: $\chi^2_{\alpha\%}(v)$ is used to indicate the critical value of χ^2 for $p = \alpha\%$ and v degrees of freedom based on the definition $\int_{\chi^2_p}^{\infty} f(\chi^2)\,d\chi^2$.

11.3. USE OF TABLES

Example 11.1. If $v = 9$, find the critical value χ^2_p of χ^2 for which
(*a*) the area under the curve to the right of χ^2_p is 0·1

(b) the area under the curve to the left of χ_p^2 is 0·05.

$$\text{(a)} \ \chi_{10\%}^2(9) = 14\cdot68,$$
$$\text{(b)} \ \chi_{95\%}^2(9) = 3\cdot33.$$

Both answers can be read directly from the tables.

Example 11.2. Find the critical value χ_p^2 if 5 per cent of the area is to the right of χ_p^2, and v takes the following values: (a) 15, (b) 21, (c) 50, (d) 84, (e) 145, (f) 250. Answers to (a), (b) and (c) can be read directly from the tables, giving (a) 25·00, (b) 32·67 and (c) 67·50.

To obtain the answer to (d) we must linearly interpolate between $\chi_p^2(80) = 101\cdot9$ and $\chi_p^2(90) = 113\cdot1$.

$$\text{The required } \chi_p^2(84) = 101\cdot9 + (113\cdot1 - 101\cdot9) \times \left(\frac{84 - 80}{90 - 80}\right)$$

$$= 101\cdot9 + 11\cdot2 \times \frac{4}{10}$$

$$= 106\cdot4$$

To obtain the answers to parts (e) and (f) we must use the formula $\sqrt{(2\chi^2)} - \sqrt{(2v - 1)} = 1\cdot645$ which gives

$$\chi^2 = \tfrac{1}{2}[1\cdot645 + \sqrt{(2v - 1)}]^2$$

(e) $\chi_p^2 = \tfrac{1}{2}(1\cdot645 + \sqrt{289})^2 = \tfrac{1}{2}(18\cdot645)^2 = 173\cdot8$

(f) $\chi_p^2 = \tfrac{1}{2}(1\cdot645 + \sqrt{499})^2 = \tfrac{1}{2}(23\cdot983)^2 = 287\cdot6$

Example 11.3. If the variable u is χ^2 distributed with $v = 14$, find χ_1^2 and χ_2^2 such that (a) $\Pr(u > \chi_2^2) = 0\cdot025$, (b) $\Pr(u < \chi_1^2) = 0\cdot01$, (c) $\Pr(\chi_1^2 \leqslant u \leqslant \chi_2^2) = 0\cdot9$.

$$\text{(a)} \ \chi_2^2 = 26\cdot12, \text{(b)} \ \chi_1^2 = 4\cdot66.$$

These two answers are read straight from the tables.

(c) There are an infinite number of answers to this part of the question depending on how we split the 10 per cent of area outside the range. The possible answers vary from $\chi_1^2 = 0$, while $\chi_2^2 = 21\cdot06$ to $\chi_1^2 = 7\cdot79$, while $\chi_2^2 = \infty$. If we put 5 per cent of the area above χ_2^2 and 5 per cent below χ_1^2 we have $\chi_1^2 = 6\cdot57$, while $\chi_2^2 = 23\cdot68$.

Exercises 11a

1. Find the critical value χ_p^2 for the following combinations:

(a) $v = 10$ and the area to the right is 1 per cent

(b) $v = 19$ and the area to the left is 0·5 per cent

(c) $v = 45$ and the area to the right is 10 per cent

(d) $v = 77$ and the area to the left is 1 per cent
(e) $v = 104$ and the area to the left is 0·075 per cent
(f) $v = 127$ and the area to the right is 0·2 per cent
(g) $v = 284$ and the area to the right is 10 per cent

2. Find : (a) $\Pr(\chi^2 \leqslant 22\cdot31)$ if $v = 15$
(b) $\Pr(\chi^2 \geqslant 200)$ if $v = 154$
(c) $\Pr(\chi^2 \leqslant 14\cdot95)$ if $v = 30$
(d) $\Pr(\chi^2 \leqslant 33\cdot11)$ if $v = 56$

11.4. TEST FOR VARIANCE

If we wish to test the null hypothesis H_0 that the variance σ^2 of a normal population is σ_0^2 we take a random sample of size n from the population. This gives an estimate $s^2 = \dfrac{\Sigma(x - \bar{x})^2}{(n - 1)}$ of σ^2. Since $\chi^2 = \dfrac{vs^2}{\sigma^2}$, if we put $v = n - 1$ we have $\chi^2 = \dfrac{\Sigma(x - \bar{x})^2}{\sigma_0^2}$; this can be tested against the tabular value $\chi^2_{\alpha\%}(n - 1)$.

Note : $v = n - 1$ because although there were originally n independent observations $x_1, x_2, \ldots x_n$ giving n degrees of freedom, calculation of $\bar{x}$ to estimate μ has used up one degree of freedom. If μ had been known s^2 would have been given by $s^2 = \Sigma(x - \mu)^2/n$ and there would have been n degrees of freedom for the test.

Example 11.4. Assuming the following data arose as a sample from a normal distribution, test at the 5 per cent level whether it supports the hypothesis $H_0 : \sigma = 2\cdot5$ against (a) the alternative $\sigma > 2\cdot5$, (b) the alternative $\sigma \neq 2\cdot5$.

$$10\cdot1, \ 20\cdot1, \ 18\cdot8, \ 15\cdot6, \ 13\cdot0, \ 18\cdot3, \ 13\cdot2, \ 10\cdot8, \ 16\cdot0$$

$$\Sigma x = 135\cdot9, \quad \Sigma x^2 = 2153\cdot59$$

$$\chi^2 = \frac{\Sigma(x - \bar{x})^2}{\sigma_0^2} = \frac{n\Sigma x^2 - (\Sigma x)^2}{n\sigma_0^2} = \frac{9 \times 2153\cdot59 - (135\cdot9)^2}{9 \times 2\cdot5^2} = 16\cdot24$$

Since χ^2 tables are single-tailed we test (a) using $\chi^2_{5\%}(8) = 15\cdot51$. This gives a significant result and we must reject the null hypothesis $\sigma = 2\cdot5$ in favour of the alternative hypothesis $\sigma > 2\cdot5$.

To test (b) we divide the 5 per cent into equal amounts at both ends of the distribution. Thus we test against $\chi^2_{97\cdot5\%}(8) = 2\cdot18$ and $\chi^2_{2\cdot5\%}(8) = 17\cdot53$. Since the value of χ^2 is within this range we cannot reject the null hypothesis $\sigma = 2\cdot5$ in favour of the alternative $\sigma \neq 2\cdot5$.

Exercises 11b

1. The following sample was obtained randomly from a normal population. Test at the 1 per cent level of significance the null hypothesis $\sigma = 1\cdot8$ against the alternative $\sigma > 1\cdot8$.

4·7	5·3	7·1	3·7	6·7	4·0	4·6	5·3	3·6	3·7
5·4	9·1	9·1	2·6	9·8	6·7	6·2	8·2	8·2	4·9
8·4	4·3	5·0	5·9	8·1	6·4	7·1	5·1	9·4	7·7
8·9	7·0	2·1	6·2	5·6	3·2	5·3	9·4	1·5	9·9

2. If in the previous problem the normal population was known to have a mean $\mu = 6\cdot0$ does this affect the result?

11.5. ADDITIVE PROPERTY OF χ^2

If χ_1^2 and χ_2^2 are independent random variates with chi-squared distributions having v_1 and v_2 degrees of freedom respectively, then $\chi^2 = \chi_1^2 + \chi_2^2$ has a chi-squared distribution with $v_1 + v_2$ degrees of freedom. This can be extended to any number of χ^2 independent random variates. This is a useful property since the values of χ^2 become more reliable with the increase of the number of degrees of freedom.

Example 11.5. An experiment was performed on three separate occasions and the resulting values of χ^2 were: 11·5 based on 7 degrees of freedom; 16·4 based on 10 degrees of freedom; 17·6 based on 11 degrees of freedom.
Test at the 5 per cent level that the null hypothesis H_0: $\sigma = \sigma_0$ (the value used to obtain the χ^2 values given) against the alternative H_1: $\sigma > \sigma_0$.

From the tables $\chi_{5\%}^2(7) = 14\cdot07$

$$\chi_{5\%}^2(10) = 18\cdot31$$
$$\chi_{5\%}^2(11) = 19\cdot68$$

Thus, taken singly, each experiment gives a non-significant result, i.e. we cannot reject H_0. Combining the three values of χ^2 we have 45·5 based on 28 degrees of freedom, and comparing this with $\chi_{5\%}^2(28) = 41\cdot34$ we obtain a significant result, and can reject H_0.

Exercises 11c

1. An experiment was carried out on four independent occasions and the following results were calculated using the experimental results and a value σ_0 of σ.

$$\chi^2(10) = 20, \; \chi^2(21) = 33\cdot5, \; \chi^2(3) = 6\cdot4, \; \chi^2(7) = 15$$

Test at the 1 per cent level the null hypothesis $H_0 : \sigma = \sigma_0$ against the alternative $\sigma > \sigma_0$.

11.6. CONFIDENCE INTERVALS FOR σ^2

The chi-squared distribution can be used to set up confidence intervals for σ^2, the variance of a normal population.

For a given sample from a normal population we can form $\chi^2 = (vs^2/\sigma^2)$ for given values of σ^2. We can also find χ_1^2 and χ_2^2 such that $\Pr(\chi_1^2 \leqslant \chi^2 \leqslant \chi_2^2) = (100 - \alpha)/100$ [see Example 11.3 (c)].

Thus for a given sample we can choose a range of values for σ^2 which will give a non-significant result, i.e. so long as $\chi_1^2 \leqslant vs^2/\sigma^2 \leqslant \chi_2^2$ we will have a non-significant result.

Assuming that α is divided equally at either end of the distribution we can obtain a $(100 - \alpha)$ per cent confidence interval for χ^2 using

$$\chi_{100-\alpha/2}^2(v) \leqslant \frac{vs^2}{\sigma^2} \leqslant \chi_{\alpha/2}^2(v)$$

which gives:

$$\frac{vs^2}{\chi_{\alpha/2}^2(v)} \leqslant \sigma^2 \leqslant \frac{vs^2}{\chi_{100-\alpha/2}^2(v)}$$

as a $(100 - \alpha)$ per cent confidence interval for σ^2. Since everything is positive we can obtain a $(100 - \alpha)$ per cent confidence interval for σ using:

$$\sqrt{\left(\frac{vs^2}{\chi_{\alpha/2}^2(v)}\right)} \leqslant \sigma \leqslant \sqrt{\left(\frac{vs^2}{\chi_{100-\alpha/2}^2(v)}\right)}$$

Example 11.6. Find a 95 per cent confidence interval for the variance of the normal population from which a random sample giving $\sum_{i=1}^{7}(x_i - \bar{x})^2 = 40$ was selected.

$$\chi_{2\frac{1}{2}\%}^2(6) = 14{\cdot}45, \quad \chi_{97{\cdot}5\%}^2(6) = 1{\cdot}24$$

$$\therefore \quad \frac{40}{14{\cdot}45} \leqslant \sigma^2 \leqslant \frac{40}{1{\cdot}24}$$

$$\therefore \quad 2{\cdot}77 \leqslant \sigma^2 \leqslant 32{\cdot}25$$

Exercises 11d

1. The estimate of the population variance σ^2 from a random sample of size 20 was 16·4. Find a 99 per cent confidence interval for the population standard deviation.

11.7. OBSERVED AND THEORETICAL FREQUENCIES

A measure of the discrepancy existing between the observed frequencies o_i and the expected frequencies e_i is given by the statistic

$$\chi^2 = \sum_{i=1}^{k} \frac{(o_i - e_i)^2}{e_i}$$

The total frequency is $N = \sum_{i=1}^{k} o_i = \sum_{i=1}^{k} e_i$

The expected frequencies are calculated using a null hypothesis, i.e. we assume a distribution for the source of the observed values and work out the expected numbers using this distribution.

The statistic χ^2 has a distribution which is approximated by the chi-squared distribution if the expected frequencies are at least equal to five (see Example 11.10). This improves as the expected frequencies become larger. The value of χ^2 obtained, using the observed and expected frequencies, can be tested by using the critical value $\chi^2_{\alpha\%}(v)$ given in the tables. A significant result implies that the assumed distribution was not a good fit.

The question now arises, what is the value of v, the number of degrees of freedom. The answer can be given in two parts.

(a) $v = k - 1$ if the expected frequencies can be computed using only the fact that $N = \sum_{i=1}^{k} o_i$.

(b) $v = k - 1 - m$ if the expected frequencies can only be computed if we estimate m parameters using the observed frequencies.

Example 11.7. Five identical coins were tossed 320 times and the observed frequencies of the number of heads per toss were as given in *Table 11.2.*

Table 11.2

No. heads	0	1	2	3	4	5
Frequency	30	60	120	80	20	10

Test if the coins are biased at the 5 per cent level. Coins tossed in this manner should follow the binomial distribution, so assuming that the coins are unbiased the probability of a head is $\frac{1}{2}$ (this is the null hypothesis). The expected frequencies are given by:

$320 \left[(\frac{1}{2})^5 + 5(\frac{1}{2})^4(\frac{1}{2}) + 10(\frac{1}{2})^3(\frac{1}{2})^2 + 10(\frac{1}{2})^2(\frac{1}{2})^3 + 5(\frac{1}{2})(\frac{1}{2})^4 + (\frac{1}{2})^5 \right]$

$\qquad\qquad = 10 + 50 + 100 + 100 + 50 + 10$

We calculate χ^2 from *Table 11.3* thus:

Table 11.3

No. heads	0	1	2	3	4	5
Obs. frequency	30	60	120	80	20	10
Exp. frequency	10	50	100	100	50	10
$\lvert o - e \rvert$	20	10	20	20	30	0
$(o - e)^2$	400	100	400	400	900	0
$\dfrac{(o - e)^2}{e}$	40	2	4	4	18	0

$$\chi^2 = \sum_{i=1}^{6} \frac{(o - e)^2}{e} = 68$$

Since the probability of a head was assumed to be $\frac{1}{2}$ and not calculated from the observed frequencies, we have $6 - 1 = 5$ degrees of freedom. $\chi^2_{5\%}(5) = 11\cdot07$. Therefore we must reject the null hypothesis that the coins are unbiased.

Note: the chi-squared test for goodness of fit is one-tailed. However, care must be exercised when a too low value of χ^2 is obtained since a certain amount of variability is always expected in an experiment. A low value of χ^2 should be tested using $\chi^2_p(v)$ for large values of p. If $\chi^2 < \chi^2_p(v)$ the conclusion would be that the observed frequencies were not unbiased and the method of obtaining the observed frequencies should be investigated.

Example 11.8. A die was rolled 102 times and the frequencies as given in *Table 11.4* were observed.

Table 11.4

Face	1	2	3	4	5	6
Frequency	19	15	16	15	18	19

Test if the die is unbiased at the 5 per cent level.
The expected frequencies are all 17

$$\therefore \lvert o - e \rvert = 2 \quad 2 \quad 1 \quad 2 \quad 1 \quad 2$$
$$(o - e)^2 = 4 \quad 4 \quad 1 \quad 4 \quad 1 \quad 4$$

$$\sum \frac{(o - e)^2}{e} = \frac{18}{17} = 1.06$$

$$\chi^2_{95\%}(5) = 1.145$$

The fit is too good and there is cause to doubt the method of rolling.

11.8. TEST FOR THE BINOMIAL DISTRIBUTION USING χ^2

Example 11.9. Example 11.7 could have asked the question: Does the data come from the binomial distribution? The null hypothesis states that the expected frequencies are given by $320(q + p)^5$ where p is the probability of a head when one coin is tossed. Since now no assumptions can be made about p we must estimate p from the data.

Mean No. of heads

$$= \frac{\Sigma fo}{\Sigma f} = \frac{30 \times 0 + 60 \times 1 + 120 \times 2 + 80 \times 3 + 20 \times 4 + 10 \times 5}{30 + 60 + 120 + 80 + 20 + 10}$$

$$= \frac{670}{320} = 2.09375 = np = 5p$$

$$\therefore p = 0.41875 \text{ and } q = 1 - p = 0.58125.$$

The expected frequencies are given by $320 (0.58125 + 0.41875)^5$, i.e.
$21.23 + 76.48 + 110.19 + 79.39 + 28.60 + 4.12$

$\|o - e\| =$ 8.77	16.48	9.81	0.61	8.60	5.88
$(o - e)^2 =$ 76.9129	271.5904	96.2361	0.3721	73.96	34.5744
$\dfrac{(o - e)^2}{e} =$ 3.623	3.551	0.873	0.005	2.586	8.392

$$\sum \frac{(o - e)^2}{e} = 19.03, \text{ we have } 6 - 1 - 1 = 4 \text{ degrees of freedom}$$

$\chi^2_{5\%}(4) = 9.49$. Therefore we can reject the null hypothesis that the distribution is a binomial one.

Exercise 11e

1. Three dice are tossed 72 times and the number of times 1 was recorded is given in *Table 11.5*.

Table 11.5

No. of 1s	0	1	2	3
Frequency	40	20	8	4

(a) Test if the probability of a 1 is a $\frac{1}{6}$ (at the 1 per cent level).

(b) Test if the data comes from a binomial distribution (at the 1 per cent level).

11.9. TEST FOR THE POISSON DISTRIBUTION USING χ^2

Example 11.10. Test if the sample given in *Table 11.6* could have come from a population which follows a Poisson distribution.

Table 11.6

x	0	1	2	3	4	5	6
Frequency	10	52	25	10	2	1	0

The null hypothesis gives the expected frequencies as $N \times P(x)$ where $N = \Sigma f$ and $P(x)$ are the Poisson probabilities $\dfrac{e^{-a}a^x}{x!}$ $\left(a = \dfrac{\Sigma fx}{\Sigma f}\right)$.

From the above data $N = 100$, $a = \dfrac{145}{100} = 1.45$

$P(0) = e^{-1.45} = 0.2349$ expected number $= 23.49$

$P(1) = 1.45 \times P(0) = 0.3406$ expected number $= 34.06$

$P(2) = \dfrac{1.45}{2} \times P(1) = 0.2469$ expected number $= 24.69$

$P(3) = \dfrac{1.45}{3} \times P(2) = 0.1193$ expected number $= 11.93$

$P(4) = \dfrac{1.45}{4} \times P(3) = \dfrac{0.0432}{0.9849}$ expected number $= 4.32$

$P(>4) = 1 - 0.9849 = 0.0151$ expected number $= 1.51$

We combine $P(4)$ and $P(>4)$ to obtain expected frequencies greater than five and obtain the results as given in *Table 11.7*.

$$\sum \frac{(o - e)^2}{e} = 18.88, \text{ degrees of freedom} = 5 - 1 - 1 = 3$$

$\chi^2_{5\%}(3) = 7.81$ therefore we reject the null hypothesis that the population follows a Poisson distribution.

Table 11.7

x	0	1	2	3	4 or over
e	23·49	34·06	24·69	11·93	5·83
o	10	52	25	10	3
$\lvert o - e \rvert$	13·49	17·94	0·31	1·93	2·83
$(o - e)^2$	181·9801	321·8436	0·0961	3·7249	8·0089
$\dfrac{(o - e)^2}{e}$	7·75	9·45	0·00	0·31	1·37

11.10. TEST FOR NORMALITY USING χ^2

The value of $\chi^2 = \Sigma[(o - e)^2/e]$ can only be calculated from data in the form of a frequency table. Thus a sample which is thought to come from a normal population must be arranged into classes to give a frequency table before we can use chi-squared to test the fit. The expected frequencies are calculated using the normal distribution with μ and σ^2 estimated by the sample $\bar{x}$ and s^2. If the data has not been grouped into classes then it is best to group it so that the expected frequencies are as near equal as possible.

Example 11.11. Test if the frequency distribution in *Table 11.8* is a sample from a normal population:

Table 11.8

Mid-interval value (x)	7·0	8·0	9·0	10·0	11·0
Class frequency	10	36	82	56	16

$$\bar{x} = \frac{\Sigma fx}{\Sigma f} = \frac{1832}{200} = 9\cdot16,\ \Sigma fx^2 = 16972$$

$$s^2 = \frac{1}{199}\left(16972 - \frac{1832^2}{200}\right) = 0\cdot9591959$$

$$\therefore \qquad s = 0\cdot9794$$

Assuming that the normal distribution holds, we use our estimated mean 9·16 and estimated standard deviation 0·9794, to give the appropriate normal distribution. The shaded areas in *Figure 11.4* will

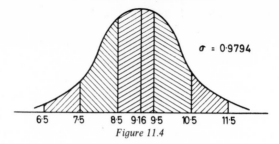

Figure 11.4

give the probability of obtaining a value in the given classes. The expected frequencies are obtained by multiplying the probabilities obtained by the total frequency.

To find the probabilities proceed as in *Table 11.9*:

Table 11.9

x	$z = \dfrac{x - \mu}{\sigma}$	*Area to right of z* $\phi(z)$	*Area between two consecutive values of x, say* x_n, x_{n+1}, i.e. $\Pr(x_n < x < x_{n+1})$	*Area × frequency*
$-\infty$	$-\infty$	1·0000		
6·5	$-2·716$	0·9967	0·0033	0·66
7·5	$-1·695$	0·9550	0·0417	8·34
8·5	$-0·684$	0·7530	0·2020	40·40
9·5	$+0·347$	0·3643	0·3887	77·74
10·5	$+1·368$	0·0856	0·2787	55·74
11·5	$+2·389$	0·00844	0·07716	15·432
$+\infty$	$+\infty$	0·00000	0·00844	1·688

The expected frequencies are as shown in the end column. The two values at each end are combined to obtain frequencies greater than five and give:

o	10	36	82	56	16
e	9	40·40	77·74	55·74	17·12
$|o - e|$	1	4·4	4·26	0·26	1·12
$(o - e)^2$	1	19·36	18·1476	0·0676	1·2544
$\dfrac{(o - e)^2}{e}$	0·111	0·4792	0·2334	0·0012	0·0733

$$\sum \frac{(o - e)^2}{e} = 0·8983$$

We have estimated μ and σ^2 from the data and therefore have $5 - 1 - 2$ = 2 degrees of freedom $\chi^2_{5\%}(2) = 5\cdot99$ and $\chi^2_{95\%}(2) = 0\cdot103$. Since $0\cdot8983$ lies between $5\cdot99$ and $0\cdot103$ we conclude that the evidence points to a normal population.

11.11. CONTINGENCY TABLES

A table of data that can be sub-totalled in two or more directions is known as a *contingency table*. Consider a two-dimensional table, if there are h rows and k columns we have an $h \times k$ contingency table. The observed frequencies can be denoted by $o_{i,j}$ $(i = 1, 2, \ldots h, j = 1, 2, \ldots k)$ and the expected frequencies, which are obtained by applying some null hypothesis to the table, by $e_{i,j}$. We can test the null hypothesis using

$$\chi^2 = \sum_{i=1}^{h} \sum_{j=1}^{k} \frac{(o_{i,j} - e_{i,j})^2}{e_{i,j}} \text{ against } \chi^2_{\alpha\%}(\nu)$$

ν is given by:

(a) $(h - 1)(k - 1)$ if no parameters have to be estimated from the observed frequencies.

(b) $(h - 1)(k - 1) - m$ if m parameters have to be estimated from the observed frequencies.

Let us suppose that rows are levels of one factor and the columns are levels of another factor. We have the *Table 11.10*.

$N_{B,i}$ are the row totals, $N_{A,j}$ are the column totals and N is the overall total.

Let $p_1, p_2, \ldots p_k$ be the probabilities associated with the levels of

Table 11.10

	A_1	A_2			A_j			A_k	
B_1	$o_{1,1}$	$o_{1,2}$	$\cdots$	$\cdots$	$o_{1,j}$	$\cdots$	$\cdots$	$o_{1,k}$	$N_{B,1}$
B_2	$o_{2,1}$	$o_{2,2}$	$\cdots$	$\cdots$	$o_{2,j}$	$\cdots$	$\cdots$	$o_{2,k}$	$N_{B,2}$
—	—	—			—			—	—
—	—	—			—			—	—
B_i	$o_{i,1}$	$o_{i,2}$	$\cdots$	$\cdots$	$o_{i,j}$	$\cdots$	$\cdots$	$o_{i,k}$	$N_{B,i}$
—	—	—			—			—	—
B_h	$o_{h,1}$	$o_{h,2}$	$\cdots$	$\cdots$	$o_{h,j}$	$\cdots$	$\cdots$	$o_{h,k}$	$N_{B,h}$
	$N_{A,1}$	$N_{A,2}$	$\cdots$	$\cdots$	$N_{A,j}$	$\cdots$	$\cdots$	$N_{A,k}$	N

A and $q_1, q_2, \ldots q_h$ the probabilities associated with the levels of B, then

$$\sum_{j=1}^{k} p_j = \sum_{i=1}^{h} q_i = 1$$

If there is no information regarding the distributions of A and B then p_j and q_i will be estimated by $N_{A,j}/N$ and $N_{B,i}/N$ respectively. If the null hypothesis states that A and B are independent the expected frequencies are given by

$$e_{i,j} = Np_jq_i = N \cdot \frac{N_{A,j}}{N} \cdot \frac{N_{B,i}}{N} = \frac{N_{A,j} \cdot N_{B,i}}{N}$$

Example 11.12. Consider the following 3×2 contingency table, *Table 11.11* and test if A and B are independent at the 5 per cent level of significance.

Table 11.11

	A_1	A_2	A_3	
B_1	25	62	13	100
B_2	19	56	20	95
	44	118	33	195

The expected values are given in *Table 11.12*.

Table 11.12

	A_1	A_2	A_3	
B_1	$\dfrac{44 \times 100}{195}$	$\dfrac{118 \times 100}{195}$	$\dfrac{33 \times 100}{195}$	100
B_2	$\dfrac{44 \times 95}{195}$	$\dfrac{118 \times 95}{195}$	$\dfrac{33 \times 95}{195}$	95
	44	118	33	195

which in *Table 11.13* gives:

Table 11.13

	A_1	A_2	A_3	
B_1	22·564	60·513	16·923	100
B_2	21·436	57·487	16·077	95
	44	118	33	195

$$\sum \frac{(o-e)^2}{e}$$

$$= \frac{(25-22\cdot564)^2}{22\cdot564} + \frac{(62-60\cdot513)^2}{60\cdot513} + \ldots + \frac{(20-16\cdot077)^2}{16\cdot077}$$

$$= 2\cdot48$$

$$v = (3-1)(2-1) = 2$$

$$\chi^2_{5\%}(2) = 5\cdot99.$$

Since $2\cdot48 < 5\cdot99$ we conclude that A and B are independent.

Example 11.13. A drug was given to one of two groups of people who were all suffering from a complaint. The numbers cured in each group are given in *Table 11.14*. Test if the drug has helped in curing the complaint.

Table 11.14

	Cured	Not cured	
Group I	19	6	25
Group II	11	14	25
	30	20	50

The expected frequencies are given in *Table 11.15*.

Table 11.15

	Cured	Not cured	
Group I	15	10	25
Group II	15	10	25
	30	20	50

using $\dfrac{N_{A,j} \cdot N_{B,i}}{N}$

$$\frac{(o-e)^2}{e} = \frac{4^2}{15} + \frac{4^2}{15} + \frac{4^2}{15} + \frac{4^2}{15}$$

$$= 5\cdot3333^{\cdot}$$

$$v = (2-1)(2-1) = 1$$

$$\chi^2_{5\%}(1) = 3\cdot84$$

Since $5.33^{.} > 3.84$ the null hypothesis that the number cured is independent of the drug is rejected at the 5 per cent level of significance. Thus we conclude that the drug has helped in curing the complaint.

11.12. YATES CORRECTION

The approximation of $\Sigma[(o - e)^2/e]$ to the chi-squared distribution is least accurate for small values of the number of degrees of freedom. It is recommended that if there is only one degree of freedom then the chi-squared distribution is approximated better by:

$$\sum_{i=1}^{k} \frac{(|o_i - e_i| - 0.5)^2}{e_i}$$

This will only make a difference to the test if the value of $\Sigma[(o - e)^2/e]$ is near to and above the critical value $\chi^2_{\alpha\%}(1)$.

Example 11.14. Re-working Example 11.13, using Yates correction, we have:

$$\sum \frac{(|o - e| - 0.5)^2}{e} = \frac{3.5^2}{15} + \frac{3.5^2}{15} + \frac{3.5^2}{10} + \frac{3.5^2}{10}$$

$$= \frac{245}{60} = 4.083^{.}$$

The result of Example 11.13 is confirmed.

EXERCISES 11

1. Find the values of:
   (a) $\chi^2_{5\%}(12)$, (b) $\chi^2_{1\%}(7)$, (c) $\chi^2_{99.5\%}(19)$, (d) $\chi^2_{95\%}(36)$, (e) $\chi^2_{2\frac{1}{2}\%}(74)$, (f) $\chi^2_{1\%}(115)$, (g) $\chi^2_{95\%}(156)$.
2. Find:
   (a) $\Pr(\chi^2 \leqslant 7.29)$ given that $v = 5$, (b) $\Pr(\chi^2 \geqslant 8.55)$ given that $v = 15$, (c) $\Pr(\chi^2 \leqslant 145)$ given that $v = 137$, (d) $\Pr(\chi^2 \leqslant 50)$ given that $v = 64$.
(Use Fisher and Yates tables).
3. Test whether the following sample came from a population with variance equal to 3.5:
27·2   28·3   25·9   26·1   28·2   29·4   26·5   25·2   26·9   27·8
Take the alternative hypothesis to be 'variance $\neq 3.5$'.
4. Set 99 per cent confidence limits for the variance of the population from which the following sample was taken:
7·4   12·5   9·7   9·0   10·6   11·3

5. A die is tossed 100 times with the results as given in *Table 11.16*.

Table 11.16

Side uppermost	1	2	3	4	5	6
Frequency	11	18	20	18	19	14

Test if the die is fair.

6. *Table 11.17* gives the frequency of occurrence of the integers 0–9 among a set of 1000 numbers:

Table 11.17

Number	0	1	2	3	4	5	6	7	8	9
Frequency	110	88	105	97	92	104	112	95	103	94

Test if the numbers are a random set.

7. In a die-tossing experiment 37 'fours' and 51 'sixes' were recorded. Is the die biased in favour of either number?

8. Prove that

$$\sum_{i=1}^{k} \frac{(o_i - e_i)^2}{e_i} = \sum_{i=1}^{k} \left(\frac{o_i^2}{e_i}\right) - N$$

9. In a survey of 1000 people, 40 per cent said they would vote Labour, 48 per cent Conservative and 12 per cent Liberal. Test the hypothesis that the true way in which the whole population will vote is 45 per cent Labour, 45 per cent Conservative and 10 per cent Liberal.

10. In question 9 what is the smallest survey which would be necessary to just reject the null hypothesis at the 5 per cent level of significance assuming the percentages remain the same?

11. Of 82 eggs from white hens, 42 are light brown, 20 are dark brown and 20 are white. Are these results consistent with the null hypothesis that the three colours should be in the ratio 5:3:2?

12. Eight coins were thrown 1024 times and the number of heads in each throw was recorded. The frequencies were as given in *Table 11.18*:

Table 11.18

No. heads	0	1	2	3	4	5	6	7	8
Frequency	4	30	113	218	286	226	116	29	2

Test if the coins are unbiased at the 5 per cent level of significance.

13. In a certain year in a given town, out of 800 road accidents 100 were serious, whereas in the following year out of 1050 accidents 105 were serious. Has the town become more susceptible to serious accidents?

14. Prove that for a 2 × 2 contingency table

a_1	b_1	N_1
a_2	b_2	N_2
N_A	N_B	N

$$\sum \frac{(o - e)^2}{e} = \frac{N(a_1 b_2 - a_2 b_1)^2}{N_1 N_2 N_A N_B}$$

15. Five hundred individuals are classified according to sex and whether or not they are short-sighted, the results are given in *Table 11.19*:

Table 11.19

	Male	Female
Not short-sighted	233	246
Short-sighted	7	14

Test if short-sightedness is independent of sex.

16. It is thought that an urn contains 100 white and 50 blue balls. Three balls are drawn at random, their colours noted and then they are replaced. *Table 11.20* gives the results of 500 such draws:

Table 11.20

3 blue	2 blue and 1 white	1 blue and 2 white	2 white
19	89	233	159

Test if the results confirm the original hypothesis of 50 blue and 100 white balls.

17. A random sample of 30 men and 70 women were asked to test if they could tell the difference between two brands of butter, the results are given in *Table 11.21*:

Table 11.21

Sex	Could tell	Couldn't tell
Male	21	9
Female	45	25

Do these results indicate that men and women differ in their ability to tell the difference between two brands of butter?

18. The number of fillings in the teeth of six-year olds is thought to follow a Poisson distribution. Test the hypothesis given in the frequency *Table 11.22*.

Table 11.22

No. of fillings (r)	0	1	2	3	4
No. of children with r fillings	60	30	8	1	1

19. *Table 11.23* gives the percentages of two samples of size 100 of English and Welsh people in regard to their mathematical ability:

Table 11.23

Nationality	Good	Fair	Bad
English	32	38	30
Welsh	53	28	19

Do these figures indicate a significant (i.e. 1 per cent level) difference between nationalities?

20. *Table 11.24* summarizes the results of a test on the hair colour of samples of women from different parts of the country. Are hair colour and place of abode related?

21. At a company which manufactures tubes for television receivers, a test of a sample batch of 1000 tubes was conducted and the number of faults in each tube were recorded, and are given in *Table 11.25*. Test if the distribution of the number of faults follows a Poisson distribution.

Table 11.24

Area	Hair Colour			
	Brown	Blonde	Auburn	Other
South	100	63	10	15
North	121	71	15	12
Midlands	114	43	19	8
Wales	85	34	21	5
Scotland	103	74	20	10

Table 11.25

No. of faults	0	1	2	3	4	5	6
Frequency	620	260	88	20	8	2	2

22. The observed frequencies of grades obtained by 200 students in Pure and Applied Mathematics are given in *Table 11.26*. Test if the performances in both subjects are related.

Table 11.26

		Pure Mathematics		
	A	16	13	10
Applied Mathematics	B	20	33	16
	C	10	28	54

13. In five colleges *A*, *B*, *C*, *D*, and *E* the number of students in Mathematics who obtained first class, second class Honours, Pass and Fail are given in *Table 11.27*.

(*a*) Carry out a chi-squared test on the data to investigate whether or not Grades and Colleges are related.

(*b*) By combining the results for colleges *A* and *B* and colleges *C*, *D* and *E*, also the results for first and second class honours, and the results for pass and fail, investigate whether or not colleges *A* and *B*

Table 11.27

College	Grade			
	First	Second	Pass	Fail
A	10	20	14	6
B	6	24	13	4
C	4	14	27	8
D	6	18	22	5
E	4	18	28	10

turn out a higher percentage of honours graduates than the other colleges.

(c) Combining the results for all the colleges, test the hypothesis that the number of students with first class honours, second class honours, pass or fail are in the ratio $1:3:3:1$.

(d) Combine the grades and test the hypothesis that the five colleges receive equal preference from students.

24. Eighty sets of ten rats were subjected to a given drug. The frequencies of the number of deaths in a set are given in *Table 11.28*.

Table 11.28

No. dead	0	1	2	3	4	5	6	7	8	9	10
Frequency	6	20	27	12	8	6	1	0	0	0	0

Can the data be assumed to follow a normal distribution? Use a 5 per cent level of significance.

25. An experimenter finds that when he counts the crystals which have formed during an experiment he has 50 large, 35 medium and 10 small. He has a theory that the ratio large: medium: small should be $6:4:1$; can he reject his theory on the results of his experiment?

26. An experimenter injects 400 rats with a drug to increase the number of births. There are 230 multiple births. In a control group of 200 rats which are not injected, there are 100 multiple births. He argues that the control group points to 200 out of 400 multiple births

if the drug has no effect and calculates

$$\chi^2 = \frac{(230 - 200)^2}{200} + \frac{(170 - 200)^2}{200}$$

$$= \frac{900}{200} + \frac{900}{200} = 9$$

$\chi^2_{1\%}(1) = 6.63$ and since $9 > 6.63$ the drug has a significant effect. Comment on his procedure.

27. The data in *Table 11.29* gives the incidence of nine mutually exclusive bone formations in the spine for 101 normal women and 52 cases of multiparous prolapse. Test whether the two groups are homogeneous as regards their distribution over various formations by finding χ^2 to three decimal places.

Table 11.29

Formations	1	2	3	4	5	6	7	8	9	Total
Normal	1	26	43	22	3	2	2	1	1	101
Prolapse	5	10	19	11	0	5	2	0	0	52

(Liv. U.)

28. The numbers below are the frequencies with which telephone calls are received at a business office in 17 successive half-hour periods throughout the day.

2, 2, 1, 3, 2, 3, 4, 3, 2, 0, 3, 4, 1, 3, 2, 3, 2.

Give reasons for supposing that these observations might follow a Poisson distribution, and test whether in fact they do. (Liv. U.)

29. The numbers of suicides in a city during a period of 200 days were recorded as in *Table 11.30*:

Table 11.30

Suicides in a day	0	1	2	3	4	5	6 and over
Number of days	91	61	30	14	3	1	0

Fit a Poisson distribution to these data and test the goodness of fit.

(Liv. U.)

12

THE F DISTRIBUTION (VARIANCE RATIO)

12.1. INTRODUCTION

IN section 9.8 it was assumed that both populations had the same variance, in order that we could use the 't' test or normal test, for the difference between two means. We now consider a distribution which allows us to use the sample variances to test if the assumption about the population variances is a reasonable one. It can also be used to test a given ratio σ_1^2/σ_2^2 of the variances and to obtain confidence limits for this ratio. The distribution is called the 'F' *distribution* in honour of R. A. Fisher who was one of the pioneers in the use of statistics.

12.2. DEFINITION

Suppose that we have two independent sample variances s_1^2 and s_2^2 which are based on v_1 and v_2 degrees of freedom respectively and are estimates of the variances σ_1^2 and σ_2^2 of two normally distributed populations. From Chapter 11 we have that $\chi_1^2 = \dfrac{v_1 s_1^2}{\sigma_1^2}$ and $\chi_2^2 = \dfrac{v_2 s_2^2}{\sigma_2^2}$ are independent chi-squared distributions.

The ratio $F = \dfrac{v_2 \chi_1^2}{v_1 \chi_2^2} = \dfrac{s_1^2 \sigma_2^2}{s_2^2 \sigma_1^2}$ follows the F distribution. The equation of the curve giving this distribution is $y = K \dfrac{F^{(v_1/2)-1}}{(v_2 + v_1 F)^{(v_1 - v_2)/2}}$ $(0 < F < \infty)$ where K is a constant depending on v_1 and v_2. *Figure 12.1* shows the curve for $v_1 > 2$. For $v_1 > 2$ the curve reaches its

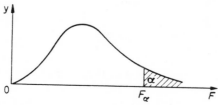

Figure 12.1. Typical F distribution curve ($v_1 > 2$)

maximum value at $F = \dfrac{v_1 - 2}{v_1} \cdot \dfrac{v_2}{v_2 + 2}$ which is always less than one, but approaches one as v_1 and v_2 become larger.

Values of $F_{\alpha\%}$, where $\alpha/100$ is the probability of obtaining a value of $F > F_{\alpha\%}$, are tabulated for various combinations of α, v_1 and v_2 (refer to the tables).

The notation used to represent the values given in the tables is $F_{\alpha\%}(v_1, v_2)$.

Example 12.1. Using the tables find the values of (a) $F_{5\%}(3,5)$, (b) $F_{2\frac{1}{2}\%}(12,21)$, (c) $F_{1\%}(8,3)$.

All these values can be read directly from the tables. For values of v_1 and v_2 not tabulated, linear interpolation is adequate.

Example 12.2. Find $F_{5\%}(16,35)$

From tables $F_{5\%}(12,30) = 2\cdot09$, $F_{5\%}(24,30) = 1\cdot89$

$$\therefore \qquad F_{5\%}(16,30) = 2\cdot09 - \frac{4}{12}(2\cdot09 - 1\cdot89) = 2\cdot02$$

$$F_{5\%}(12,40) = 2\cdot00, \; F_{5\%}(24,40) = 1\cdot79$$

$$\therefore \qquad F_{5\%}(16,40) = 2\cdot00 - \frac{4}{12}(2\cdot00 - 1\cdot79) = 1\cdot93$$

$$\therefore \qquad F_{5\%}(16,35) = 2\cdot02 - \frac{5}{10}(2\cdot02 - 1\cdot93) = 1\cdot975$$

Calculation of $F_{(100-\alpha)\%}(v_1, v_2)$. (where α is a value given in the tables).

By substitution in the formula for the F distribution it can be shown that

$$F_{(100-\alpha)\%}(v_1, v_2) = \frac{1}{F_{\alpha\%}(v_2, v_1)} \qquad \dots (12.1)$$

Example 12.3. Using tables find the value of: (a) $F_{95\%}(4,6)$, (b) $F_{97\frac{1}{2}\%}(8,7)$, (c) $F_{99\%}(5,5)$.

$$(a) \; F_{95\%}(4,6) = \frac{1}{F_{5\%}(6,4)} = \frac{1}{6\cdot16} = 0\cdot162$$

$$(b) \; F_{97\frac{1}{2}\%}(8,7) = \frac{1}{F_{2\frac{1}{2}\%}(7,8)} = \frac{1}{4\cdot53} = 0\cdot22$$

$$(c) \; F_{99\%}(5,5) = \frac{1}{F_{1\%}(5,5)} = \frac{1}{10\cdot97} = 0\cdot091$$

12.3. TESTING FOR THE EQUALITY OF TWO POPULATION VARIANCES

If we have two random samples with variances s_1^2 and s_2^2 based on v_1 and v_2 degrees of freedom we can test at a given level of significance using the F test whether there is any difference between the variances σ_1^2 and σ_2^2 of the populations from which the samples were drawn. The null hypothesis is $\sigma_1^2 = \sigma_2^2$ and the alternative hypothesis is $\sigma_1^2 \neq \sigma_2^2$. This is a two-tailed test and if we are testing at the α per cent level of significance we must test if $F = \dfrac{s_1^2}{s_2^2}$ lies within the range $F_{(100-\alpha/2)\%}(v_1, v_2)$ and $F_{\alpha/2\%}(v_1, v_2)$.

From 12.1 $F_{(100-\alpha/2)\%}(v_1, v_2) = \dfrac{1}{F_{\alpha/2\%}(v_2, v_1)}$; it is sufficient to form $F = \dfrac{\text{larger } s^2}{\text{smaller } s^2}$ and to test if $F < F_{\alpha/2\%}(v_l, v_s)$ where v_l is the number of degrees of freedom associated with the largest s^2.

Example 12.4. Two samples of size 13 and 16 give variances of 45 and 33 respectively. Test at the 5 per cent level whether the variances of the populations from which the samples were drawn are different. $F = 45/33 = 1\cdot36$. The numbers of degrees of freedom are 12 and 15. Thus we test F against $F_{2\frac{1}{2}\%}(12,15) = 2\cdot96$ which value is obtained from the tables. Since $1\cdot36 < 2\cdot96$ there is no evidence at the 5 per cent level to indicate that the population variances are different.

Example 12.5. Given that a variance of 120 based on two degrees of freedom is an estimate of $\sigma_2^2 + 3\sigma_1^2$ and another variance of 15 based on 12 degrees of freedom is an estimate of σ_2^2 test the hypothesis $\sigma_1^2 = 0$.

On the null hypothesis both 120 and 15 are estimates of σ_2^2. The alternative hypothesis is $\sigma_1^2 > 0$ since we cannot have negative variances. Therefore the test is single-sided and we test $F = 120/15 = 8$ against $F_{\alpha\%}(2,12)$, $F_{5\%}(2,12) = 3\cdot89$, $F_{1\%}(2,12) = 6\cdot93$, $F_{0\cdot1\%}(2,12) = 12\cdot97$. Thus the test gives a significant result at the 5 and 1 per cent levels but a non-significant result at the 0·1 per cent level.

12.4. CONFIDENCE LIMITS FOR THE VARIANCE RATIO $\dfrac{\sigma_1^2}{\sigma_2^2}$

If we have two sample variances s_1^2 and s_2^2 based on v_1 and v_2 degrees of freedom and two population variances σ_1^2 and σ_2^2 we can calculate $F = s_1^2\sigma_2^2/s_2^2\sigma_1^2$ and test the null hypothesis that the ratio of the population variances is σ_1^2/σ_2^2. For a two-tailed test this would involve seeing if F was in the range $F_{(100-\alpha/2)\%}(v_1, v_2)$ to $F_{\alpha/2\%}(v_1, v_2)$, i.e. we see if

$$F_{(100-\alpha/2)\%}(v_1, v_2) < \frac{s_1^2 \sigma_2^2}{s_2^2 \sigma_1^2} < F_{\alpha/2\%}(v_1, v_2)$$

If this inequality is satisfied we have a non-significant result. It is easily seen that for given values of s_1^2, s_2^2, v_1, v_2 and α there is a range of values for the ratio σ_1^2/σ_2^2 which will give a non-significant result. This range of values is a $(100 - \alpha)$ confidence interval for σ_1^2/σ_2^2 based on the samples. Rearrangement of the inequality gives

$$\frac{s_1^2}{s_2^2 F_{\alpha/2\%}(v_1, v_2)} < \frac{\sigma_1^2}{\sigma_2^2} < \frac{s_1^2}{s_2^2 F_{(100-\alpha/2)\%}(v_1, v_2)}$$

Using equation (12.1) we have that

$$\frac{s_1^2}{s_2^2 F_{\alpha/2\%}(v_1, v_2)} < \frac{\sigma_1^2}{\sigma_2^2} < \frac{s_1^2 F_{\alpha/2\%}(v_2, v_1)}{s_2^2} \qquad \dots (12.2)$$

Example 12.6. Given two samples size $n_1 = 9$ and $n_2 = 8$ with variances $s_1^2 = 42$ and $s_2^2 = 28$ respectively find a 95 per cent confidence interval for the ratio σ_1^2/σ_2^2.

Using equation (12.2) the 95 per cent confidence interval is given by

$$\frac{42}{28 F_{2\frac{1}{2}\%}(8,7)} < \frac{\sigma_1^2}{\sigma_2^2} < \frac{42 F_{2\frac{1}{2}\%}(7,8)}{28}$$

$$\frac{42}{28 \times 4\cdot90} < \frac{\sigma_1^2}{\sigma_2^2} < \frac{42 \times 4\cdot53}{28}$$

$$0\cdot31 < \frac{\sigma_1^2}{\sigma_2^2} < 6\cdot8$$

Example 12.7. Given the following two samples from two different populations find a 90 per cent confidence interval for the ratio of the population variances.

Sample I   14   16   16   15   18   14   13   12   11   18

Sample II 28·0 27·2 27·5 27·3 26·8 26·9 27·1 27·3

$$\Sigma_1 x = 147, \Sigma_1 x^2 = 2211, \Sigma_2 x = 218\cdot1, \Sigma_2 x^2 = 5946\cdot93$$

$$s_1^2 = \frac{1}{9}\left(2211 - \frac{147^2}{10}\right), \quad s_2^2 = \frac{1}{7}\left(5946\cdot93 - \frac{218\cdot1^2}{8}\right)$$

$$= \frac{2211 \times 10 - 147^2}{9 \times 10} \qquad = \frac{8 \times 5946\cdot93 - 218\cdot1^2}{7 \times 8}$$

$$= 5\cdot56 \qquad\qquad\qquad = 0\cdot14$$

therefore a 90 per cent confidence interval is given by:

$$\frac{5.56}{0.14\,F_{5\%}(9,7)} < \frac{\sigma_1^2}{\sigma_2^2} < \frac{5.56\,F_{5\%}(7,9)}{0.14}$$

$$\frac{5.56}{0.14 \times 3.68} < \frac{\sigma_1^2}{\sigma_2^2} < \frac{5.56 \times 3.29}{0.14}$$

$$10.8 \quad < \frac{\sigma_1^2}{\sigma_2^2} < \quad 130.8$$

Note: we can obtain a 90 per cent confidence interval for $\dfrac{\sigma_1}{\sigma_2}$ by taking the square root of both sides,

i.e. $$3.286 < \frac{\sigma_1}{\sigma_2} < 11.43$$

EXERCISES 12

1. Use tables to find the following: (a) $F_{1\%}(5,9)$, (b) $F_{2\frac{1}{2}\%}(7,9)$, (c) $F_{5\%}(24,12)$, (d) $F_{95\%}(4,6)$, (e) $F_{99\%}(10,12)$, (f) $F_{99.9\%}(35,10)$.

2. By finding the tabular values show that:
(a) $F_{5\%}(1,10) = t_{5\%}^2(10)$, (b) $F_{1\%}(1,12) = t_{1\%}^2(12)$.

3. From a sample size 16 taken from a population A the estimate of the variance of the population A was 6·2, while a sample of size 21 from a population B estimated the variance of population B as 3·9. Test at the 5 per cent level of significance whether the populations have the same variance. What assumption must you make?

4. The results of two independent experiments were:

A	28	16	10	15	18	23	31	22
B	104	89	110	94	86	104		

If the differences in the results are due to random errors only, test if the variance of the population of errors is the same for each experiment.

5. For Example 12.4 find a 95 per cent confidence interval for the ratio of the variances of the population errors.

6. Given $s_1^2 = 40$ and $s_2^2 = 10$, each based on 10 degrees of freedom, determine the probability that sample variances *as* divergent, as these could be estimates of the same population variance with an alternative hypothesis $\sigma_1^2 \neq \sigma_2^2$.

7. Twelve pieces of material were sampled randomly, one was lost, five were subjected to a first treatment A and six were subjected to a second treatment B. Certain test measurements were made and these had the following variances: 0·00045 with the treatment A, 0·00039 with treatment B. Find a 95 per cent confidence interval for the ratio of the population variances.

13

BIVARIATE DISTRIBUTIONS

13.1. INTRODUCTION

So far in dealing with frequency distributions we have considered only one variate, that is, we have had a univariate distribution. In this chapter populations with two variates, that is *bivariate distributions* are considered.

Suppose that we have a random sample of n pairs of values (x_i, y_i) and it is thought that there is a linear relationship $y_i = \beta_0 + \beta_1 x_i$ between x_i and y_i. The problem is to estimate the values of β_0 and β_1 using the sample data. The choice of the method of estimation is quite arbitrary, but here consideration is only given to the *method of least squares.* To apply this method of estimation we must make several assumptions, which are:

(a) x is assumed to be measured with negligible error, that is $y_i = \beta_0 + \beta_1 x_i + e_i$ gives the linear relationship allowing for errors e_i in the y_i's.

(b) The mean of the e_i's is assumed to be zero.

(c) The variances of the distributions of e_i's are assumed to be all equal to σ^2.

(d) The covariance of e_i and e_j is assumed zero $(i \neq j)$.

A further assumption that the e_i's are all normally distributed is added later when tests of significance are applied and confidence intervals found. This assumption of normality makes e_i and e_j independent $(i \neq j)$.

The method of least squares minimizes the sum of squares of the errors e_i. To do this we minimize

$$Q = \sum_{i=1}^{n} e_i^2 = \sum_{i=1}^{n} (y_i - \beta_0 - \beta_1 x_i)^2$$

with respect to β_0 and β_1. This is done using Taylor's theorem.

$$\frac{\partial Q}{\partial \beta_0} = -2 \sum_{i=1}^{n} (y_i - \beta_0 - \beta_1 x_i)$$

$$\frac{\partial Q}{\partial \beta_1} = -2 \sum_{i=1}^{n} (y_i - \beta_0 - \beta_1 x_i) x_i$$

$$\frac{\partial^2 Q}{\partial \beta_0^2} = -2 \sum_{i=1}^{n} (-1) = 2n > 0 \qquad \ldots (i)$$

$$\frac{\partial^2 Q}{\partial \beta_1^2} = -2 \sum_{i=1}^{n} (-x_i^2) = 2 \sum_{i=1}^{n} x_i^2 > 0 \qquad \ldots (ii)$$

$$\frac{\partial^2 Q}{\partial \beta_0 \partial \beta_1} = -2 \sum_{i=1}^{n} (-x_i) = 2 \sum_{i=1}^{n} x_i$$

$$\frac{\partial^2 Q}{\partial \beta_0^2} \cdot \frac{\partial^2 Q}{\partial \beta_1^2} - \left(\frac{\partial^2 Q}{\partial \beta_0 \partial \beta_1} \right)^2 = 2n \times 2 \sum_{i=1}^{n} x_i^2 - 4(\sum x_i)^2$$

$$= 4n \left(\sum x^2 - \frac{(\sum x)^2}{n} \right) = 4n \sum_{i=1}^{n} (x_i - \bar{x})^2 > 0 \ldots (iii)$$

Equations (i), (ii) and (iii) show that any solution of the two equations $\frac{\partial Q}{\partial \beta_0} = 0$ and $\frac{\partial Q}{\partial \beta_1} = 0$ gives a maximum value to Q.

Putting $\tilde{\beta}_0$ and $\tilde{\beta}_1$ as the estimates of β_0 and β_1 we find that

$$\frac{\partial Q}{\partial \beta_0} = 0 \text{ and } \frac{\partial Q}{\partial \beta_1} = 0 \text{ give}$$

$$\left. \begin{array}{c} \displaystyle\sum_{i=1}^{n} y_i = \tilde{\beta}_0 n + \tilde{\beta}_1 \sum_{i=1}^{n} x_i \\[4mm] \displaystyle\sum_{i=1}^{n} y_i x_i = \tilde{\beta}_0 \sum_{i=1}^{n} x_i + \tilde{\beta}_1 \sum_{i=1}^{n} x_i^2 \end{array} \right\} \qquad \ldots (13.1)$$

These are known as the *normal equations*, and solving, we have that (omitting the subscripts)

$$\tilde{\beta}_1 = \frac{n \sum yx - (\sum y)(\sum x)}{n \sum x^2 - (\sum x)^2}$$

$$\tilde{\beta}_0 = \bar{y} - \beta_1 \bar{x}_1 = \frac{\sum x^2 \sum y - \sum x \sum xy}{n \sum x^2 - (\sum x)^2}$$

Thus $$Y = \bar{y} - \tilde{\beta}_1 \bar{x} + \tilde{\beta}_1 x$$

or $$Y - \bar{y} = \tilde{\beta}_1 (x - \bar{x}) \qquad \ldots (13.2)$$

gives the estimated value Y of y for a given value of x.

It is left to the reader to show that $\tilde{\beta}_1$ can also be written as

$$\frac{\sum (y - \bar{y})(x - \bar{x})}{\sum (x - \bar{x})^2} = \frac{\sum (y - \bar{y})(x - \bar{x})}{(n-1)} \times \frac{(n-1)}{\sum (x - \bar{x})^2} \ldots (13.3)$$

$\frac{\sum(y - \bar{y})(x - \bar{x})}{n - 1}$ is an estimate of the *covariance* (see Chapter 14)

of x and y and is denoted by $s_{x,y}$. $\frac{\sum(x - \bar{x})^2}{n - 1}$ is the estimate of the

variance of x which we write as s_x^2.

Thus $\tilde{\beta}_1 = \frac{s_{x,y}}{s_x^2}$ and the equation becomes

$$Y - \bar{y} = \frac{s_{x,y}}{s_x^2}(x - \bar{x}) \qquad \ldots (13.4)$$

Note: this equation passes through the point $(\bar{x}, \bar{y})$.

Thus we have that the least squares estimate of Q is $\sum(y - Y)^2$

$$= \sum[(y - \bar{y}) - (Y - \bar{y})]^2$$
$$= \sum(y - \bar{y})^2 - 2\sum(y - \bar{y})(Y - \bar{y}) + \sum(Y - \bar{y})^2 \qquad \ldots (i)$$

However, since $Y - \bar{y} = \tilde{\beta}_1(x - \bar{x})$ and $\beta_1 = \frac{\sum(y - \bar{y})(x - \bar{x})}{\sum(x - \bar{x})^2}$.

the centre term in equation (i) becomes

$$-2\sum(y - \bar{y})(Y - \bar{y}) = -2\sum(y - \bar{y})\tilde{\beta}_1(x - \bar{x}) = -2\tilde{\beta}_1^2\sum(x - \bar{x})^2$$

$$= -2\sum[\tilde{\beta}_1(x - \bar{x})]^2$$

$$= -2\sum(Y - \bar{y})^2 \qquad \ldots (ii)$$

Substituting result (ii) in (i) we have

$$Q = \sum(y - \bar{y})^2 - \sum(Y - \bar{y})^2 \qquad \ldots (13.5)$$

$\sum(y - \bar{y})^2$ is called the corrected total sum of squares, $\sum(Y - \bar{y})^2$ is called the sum of squares due to regression and is calculated using:

$$\tilde{\beta}_1^2\sum(x - \bar{x})^2 \text{ or } \tilde{\beta}_1\sum(y - \bar{y})(x - \bar{x}) \text{ or } \tilde{\beta}_1\left[\frac{n\sum xy - (\sum x)(\sum y)}{n}\right] \ldots (13.6)$$

Thus Q, which is called the sum of squares about regression, is equal to the corrected total sum of squares minus the sum of squares due to regression. It is based on $(n - 2)$ degrees of freedom since the observed values y_i have been used to calculate two parameters $\bar{y}$ and $\tilde{\beta}_1$.

The estimate of σ^2 is $\qquad \frac{Q}{(n - 2)} = s_E^2 \qquad \ldots (13.7)$

The variance of β_1 is $\qquad \frac{\sigma^2}{\sum(x - \bar{x})^2} \qquad \ldots (13.8)$

Equation (13.8) is proved as follows:

$$\text{var}(\beta_1) = \text{var}\left[\frac{\sum(x - \bar{x})(y - \bar{y})}{\sum(x - \bar{x})^2}\right]$$

$$= \text{var}\left[\frac{\sum(x - \bar{x})y - \sum(x - \bar{x})\bar{y}}{\sum(x - \bar{x})^2}\right]$$

$$= \text{var}\frac{\sum(x - \bar{x})y}{\sum(x - \bar{x})^2}\left[\text{since } \sum(x - \bar{x})\bar{y} = \bar{y}\sum(x - \bar{x}) = 0\right]$$

$$= \frac{1}{[\sum(x - \bar{x})^2]^2}\,\text{var}\left[\sum(x - \bar{x})y\right]$$

$$= \frac{1}{[\sum(x - \bar{x})^2]^2}\,\sum\left[\text{var}(x - \bar{x})y\right]$$

$$= \frac{1}{[\sum(x - \bar{x})^2]^2}\,\sum(x - \bar{x})^2\,\text{var}(y)$$

$$= \frac{1}{\sum(x - \bar{x})^2} \times \text{var}(y)$$

$$= \frac{1}{\sum(x - \bar{x})^2} \times \sigma^2$$

This is estimated by
$$\frac{s_E^2}{\sum(x - \bar{x})^2}.$$

In order to test the null hypothesis $\beta_1 = 0$ we use the t test and calculate:

$$t = \frac{\tilde{\beta}_1 - 0}{\sqrt{(\text{var }\beta_1)}} = \frac{\tilde{\beta}_1}{s_E/\sqrt{[\sum(x - \bar{x})^2]}}$$

$$= \frac{\tilde{\beta}_1\sqrt{[\sum(x - \bar{x})^2]}}{s_E} = \sqrt{\left(\frac{\beta_1^2\sum(x - \bar{x})^2}{s_E^2}\right)}$$

$$= \sqrt{\left[\frac{\text{Sum of squares due to regression} \times (n - 2)}{\text{Sum of squares about regression}}\right]} \quad \ldots (13.9)$$

this is tested against $t_{\alpha\%}(n - 2)$.

Example 13.1. Fit and test the linear regression line for the data given in *Table 13.1.*

Table 13.1

x	0	1	2	3	4
y	12	12·5	13·1	13·8	14·5

$$\sum x = 10, \sum x^2 = 30, \sum y = 65\cdot9, \sum y^2 = 872\cdot55, \sum xy = 138\cdot1$$

$$\therefore \qquad \bar{y} = \frac{65\cdot9}{5} = 13\cdot18, \quad \bar{x} = \frac{10}{5} = 2$$

$$\tilde{\beta}_1 = \frac{n\sum xy - (\sum x)(\sum y)}{n\sum x^2 - (\sum x)^2} = \frac{5 \times 138\cdot1 - 10 \times 65\cdot9}{5 \times 30 - 100}$$

$$= 0\cdot63$$

Using equation (13.2) the equation of the regression line is

$$y = 13\cdot18 + 0\cdot63\,(x - 2)$$
$$= 0\cdot63x + 11\cdot92$$

To test $\beta_1 = 0$ we proceed as follows:

Total sum of squares $= \sum y^2 - \dfrac{(\sum y)^2}{n} = 872\cdot55 - \dfrac{65\cdot9^2}{5} = 3\cdot988$

Sum of squares due to regression is given by

$$\tilde{\beta}_1 \left[\frac{n\sum xy - (\sum x)(\sum y)}{n} \right] = 0\cdot63 \times \frac{31\cdot5}{5} = 3\cdot969$$

Thus the sum of squares about regression $= 3\cdot988 - 3\cdot969 = 0\cdot019$. To test $\beta_1 = 0$ we form:

$$t = \sqrt{\left(\frac{3\cdot969}{0\cdot019} \times \frac{3}{1} \right)} = \sqrt{626\cdot7} \simeq 25$$

$t_{1\%}(3) = 5\cdot841$, since $25 \gg 5\cdot841$ the linear regression is highly significant.

13.2. CONFIDENCE INTERVALS FOR β_1 AND β_0

Having estimated β_1 and β_0, using the sample data, we can increase the value of the results if confidence intervals can be found for β_1 and β_0. These are easily found if we know the variances of $\tilde{\beta}_1$ and $\tilde{\beta}_0$.

$$\text{var}\tilde{\beta}_1 = \frac{\sigma^2}{\sum(x - \bar{x})^2} \quad [\text{see equation (13.8)}]$$

$$\text{var}\,\tilde{\beta}_0 = \text{var}\,(\bar{y} - \tilde{\beta}_1\bar{x}) = \text{var}\,\bar{y} + \bar{x}^2\,\text{var}\,\tilde{\beta}_1$$

since $\bar{y}$ and $\tilde{\beta}_1$ are statistically independent (the proof of this is beyond the scope of this book) therefore

$$\text{var}\,\tilde{\beta}_0 = \frac{\sigma^2}{n} + \frac{\bar{x}^2\sigma^2}{\sum(x - \bar{x})^2}$$

If σ^2 is unknown we estimate it by s_E^2 and the $(100 - \alpha)$ per cent confidence limits are:

(a) for β_1
$$\tilde{\beta}_1 \pm t_\alpha \frac{s_E}{\sqrt{\sum(x - \bar{x})^2}}$$

(b) for β_0
$$\bar{y} - \tilde{\beta}_1\bar{x} \pm t_\alpha \cdot s_E \cdot \sqrt{\left[\frac{1}{n} + \frac{\bar{x}^2}{\sum(x - \bar{x})^2}\right]}$$

Example 14.2. Using the data of Example 13.1 we find a 95 per cent confidence interval for β_1 is

$$0.63 \pm t_{5\%}(3)\sqrt{\frac{0.019}{10}} = 0.63 \pm 3.18\sqrt{0.0019}$$

$$= 0.63 \pm 3.18 \times 0.0436 = 0.491 - 0.769$$

and for β_0 is

$$11.92 \pm t_{5\%}(3)\sqrt{\left[0.019\left(\frac{1}{5} + \frac{4}{10}\right)\right]} = 11.92 \pm 0.3396$$

$$= 11.58 - 12.26.$$

13.3. CORRELATION

Suppose we have a problem in which we do not wish to estimate one variable from another and we cannot assume one variable is more likely than the other to contain the error. However, we are interested in any association between the variables, that is, we are interested in *interdependence* and not in *dependence*.

As an example consider a hat manufacturer. From a sample of the population he could take measurements of the pairs, head length and head breadth. His interest would be to ascertain if these were associated and not in predicting one from the other so that he would not manufacture hat sizes which he could not sell. The data he obtained from his sample when plotted on a graph could have appeared as on one of the three graphs shown in *Figure 13.1* or as a graph somewhere between the three.

We say that graph (*i*) shows a positive correlation, (*ii*) shows zero correlation and (*iii*) shows negative correlation.

The problem is to find a measure of the correlation for the actual data and test whether it is significantly different from some null hypothesis value (generally zero).

The linear regression equations of y on x and x on y can be written

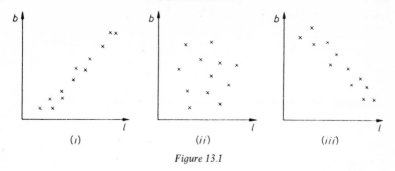

Figure 13.1

respectively:

$$y - \bar{y} = \frac{s_{x,y}}{s_x^2}(x - \bar{x})$$

$$x - \bar{x} = \frac{s_{x,y}}{s_y^2}(y - \bar{y})$$

These equations may be rearranged to give:

$$\frac{y - \bar{y}}{s_y} = \frac{s_{x,y}}{s_x s_y}\left(\frac{x - \bar{x}}{s_x}\right)$$

and

$$\frac{x - \bar{x}}{s_x} = \frac{s_{x,y}}{s_x s_y}\left(\frac{y - \bar{y}}{s_y}\right)$$

we let

$$r = \frac{s_{x,y}}{s_x s_y} \qquad \dots (13.10)$$

r is known as the *product moment correlation coefficient*. For $r = +1$ or $r = -1$ both equations are identical and we have perfect positive or negative correlation respectively.

By the Cauchy–Schwarz inequality

$$s_{x,y}^2 = \left[\frac{\sum(x - \bar{x})(y - \bar{y})}{n - 1}\right]^2 \leqslant \frac{\sum(x - \bar{x})^2}{(n - 1)}\frac{\sum(y - \bar{y})^2}{(n - 1)} = s_x^2 s_y^2$$

$\therefore \qquad \dfrac{s_{x,y}^2}{s_x^2 s_y^2} \leqslant 1 \qquad$ that is $\qquad r^2 \leqslant 1$

$\therefore \qquad -1 \leqslant r \leqslant 1$

If we denote by β_x the regression coefficient of y on x, and by β_y the regression coefficient of x on y, then

$$r^2 = \frac{s_{x,y}^2}{s_x^2 s_y^2} = \frac{s_{x,y}}{s_x^2} \times \frac{s_{x,y}}{s_y^2} = \beta_x \cdot \beta_y$$

$$\therefore \qquad r = \sqrt{(\beta_x \cdot \beta_y)} \text{ and takes the sign of } s_{x,y}.$$

Example 13.3. Find the product moment correlation coefficient for the pairs of data in *Table 13.2.*

Table 13.2

Vehicles (x) (millions)	2·6	3·1	3·5	3·7	4·1	4·4	4·6	4·9	5·3	5·8
Casualties (y) (thousands)	137	160	166	155	177	201	216	212	225	237

We can take a false origin and remove any common factors from the values of both x and y, i.e. $D = \dfrac{x - A}{C}$ and $F = \dfrac{y - B}{E}$

The equation for $r = \dfrac{s_{x,y}}{s_x s_y} = \sqrt{\left(\dfrac{(n\sum xy - \sum x \sum y)^2}{(n\sum x^2 - (\sum x)^2)(n\sum y^2 - (\sum y)^2)}\right)}$

becomes $\qquad r = \sqrt{\left(\dfrac{(n\sum DF - \sum D \sum F)^2}{(n\sum D^2 - (\sum D)^2)(n\sum F^2 - (\sum F)^2)}\right)}$

x and y are replaced by D and F without changing the formula, i.e. the formula for r is invariant to changes in origin and unit.

In the example (*Table 13.3*) we take $A = 0$, $C = 0.1$, $B = 130$, and $E = 1$.

Table 13.3

$D = \dfrac{x}{0.1}$	26	31	35	37	41	44	46	49	53	58
$F = y - 130$	7	30	36	25	47	71	86	82	95	107

$$\sum D = 420, \sum D^2 = 18{,}538, \sum F = 586, \sum F^2 = 44{,}714, \sum DF = 27{,}563$$

$$\therefore \qquad r^2 = \frac{(10 \times 27{,}563 - 420 \times 586)^2}{(10 \times 18{,}538 - 420^2)(10 \times 44{,}714 - 586^2)}$$

$$\doteq \frac{29{,}510^2}{8980 \times 103{,}744}$$

$$= 0\cdot93475$$

$$r = 0\cdot965$$

Having obtained a value for r we need to test it for significance. There are three cases to be considered; we can test the null hypothesis that the true correlation is either (a) zero, (b) $\rho = \rho_0$ (ρ is the population correlation coefficient), (c) the true correlation ρ_1 associated with r_1 is the same as a correlation ρ_2 associated with r_2.
The tests are based on the assumption of normal distributions.

(a) Perfect correlation gives linear regression and we can therefore use the same test as for linear regression, thus we test

$$t = \sqrt{\left[\frac{\text{sum of squares due to regression}}{\text{sum of squares about regression}} \times \frac{(n-2)}{1}\right]}$$

[see equation (13.9)]

$$= \sqrt{\left(\frac{r^2(n-2)}{(1-r^2)}\right)} \tag{13.10}$$

(the proof of this step is left to the reader)

against $t_{\alpha\%}(n-2)$.

(b) Fisher's z transformation is related to the correlation coefficient by the equation:

$$z = \frac{1}{2}\log_e\left(\frac{1+r}{1-r}\right) = \text{arc tanh } r$$

and is normally distributed for normal variates with mean

$$\mathscr{E}(z) \doteq \frac{1}{2}\log_e\left(\frac{1+\rho}{1-\rho}\right) \text{ and variance}$$

$$\text{var } z \doteq \frac{1}{n-3}$$

Thus we test:

$$t = \frac{z - \mathscr{E}(z)}{\sqrt{\text{var } z}} = [z - \mathscr{E}(z)] \times \sqrt{(n-3)}$$

against the normal distribution.

(c) We calculate z_1 and z_2 using r_1 and r_2 respectively. $z_1 - z_2$ will be approximately normally distributed with mean $\rho_1 - \rho_2$ and variance $\dfrac{1}{(n_1 - 3)} + \dfrac{1}{(n_2 - 3)}$ (n_1 and n_2 are the two sample sizes).

Under the null hypothesis of equal correlation $\rho_1 = \rho_2$ we test

$$t = \frac{z_1 - z_2}{\sqrt{\left(\dfrac{1}{n_1 - 3} + \dfrac{1}{n_2 - 3}\right)}}$$

against the normal distribution.

Example 13.4. For the data of Example 13.3 test the null hypotheses
(a) $\rho = 0$ and (b) $\rho = 0.7$.

$$(a) \qquad t \quad = \sqrt{\frac{0.93475 \times 8}{0.06525}} = \sqrt{114.605} = 10.7$$

$$t_{5\%}(8) = 2.31, \, t_{1\%}(8) = 3.36$$

Since $10.7 > 3.36$ the result is significant at the 1 per cent level and we reject the null hypothesis $\rho = 0$.

$$(b) \qquad z \quad = \frac{1}{2}\log_e\left(\frac{1 + 0.965}{1 - 0.965}\right) = \frac{1}{2}\log_e\left(\frac{1.965}{0.035}\right)$$

$$= 2.014$$

$$\mathscr{E}(z) = \frac{1}{2}\log_e\left(\frac{1 + 0.7}{1 - 0.7}\right) = 0.867$$

$$t \quad = (2.014 - 0.867)\sqrt{7}$$

$$= 3.034.$$

The two-tailed 1 per cent critical value of z is 2.58; since $3.034 > 2.58$ the test is significant at the 1 per cent level and we reject the null hypothesis that $\rho = 0.7$.

13.4. GROUPED DATA

If there are several combinations of the same x and y (this will occur most frequently if the data are grouped) then the data is best presented as a two-way *contingency table*.

Example 13.5. Table 13.4 gives the frequency of occurrence of the various groups of the pairs mathematics mark and physics mark. Calculate the product moment correlation coefficient.

Table 13.4

	Centre of interval	Mathematics mark (x)				
		44·5	54·5	64·5	74·5	84·5
	44·5	2	4			
Physics	54·5		5	6		
mark	64·5		2	1	4	
(y)	74·5			7	1	3
	84·5					1

The figures given in each cell of the table are the frequencies of the occurrence of the given combinations represented by the cell. The working is simplified if we use a false origin and a unit for each of x and y. Putting $u = (x - A)/C$ and $v = (y - B)/E$ will not affect the calculation of r (see also Example 13.3)

$$r = \frac{n \sum (f_{x,y} uv) - [\sum (f_x u)][\sum (f_y v)]}{\sqrt{[n \sum (f_x u^2) - (\sum (f_x u))^2][n \sum (f_y v^2) - (\sum (f_y v))^2]}}$$

where $f_{x,y}$ is the individual cell frequency, f_x is a column total, and f_y is a row total. Putting $u = \dfrac{x - 64·5}{10}$ and $v = \dfrac{y - 64·5}{10}$ we have:

Table 13.5. Mathematics mark (x)

v \ u	−2	−1	0	1	2	f_y
−2	2 ₄	4 ₂				6
−1		5 ₁	6 ₀			11
0		2 ₀	1 ₀	4 ₀		7
1			7 ₀	1 ₁	3 ₂	11
2					1 ₄	1
f_x	2	11	14	5	4	36

The computation of $\sum (f_{x,y} uv)$ is helped if the product uv is placed in the corner of each cell containing a frequency (see Table 13.5).

$$\sum(f_{x,y}uv) = 32, \sum(f_x u) = -2, \sum(f_x u^2) = 40,$$
$$\sum(f_y v) = -21, \sum(f_y v^2) = 50$$

$$r^2 = \frac{[36 \times 32 - (-21)(-2)]^2}{[36 \times 40 - (-2)^2][36 \times 50 - (-21)^2]}$$

$$= \frac{1,232,100}{1,951,524}$$

$$\therefore \qquad r = 0.795.$$

We can find the regression equations of y on x and x on y for data presented as in Example 13.5 by using:

$$y - \bar{y} = \sqrt{\left\{ \frac{n\sum(f_{x,y}xy) - [\sum(f_x x)][\sum(f_y y)]}{n\sum(f_x x^2) - [\sum(f_x)x)]^2} \right\}} (x - \bar{x})$$

$$x - \bar{x} = \sqrt{\left[\frac{n\sum(f_{x,y}xy) - [\sum(f_x x)][\sum(f_y y)]}{n\sum(f_y y^2) - [\sum(f_y y)]^2} \right]} (y - \bar{y})$$

13.5. RANK CORRELATION

Rank correlation can be used to give a quick approximation to the product moment correlation coefficient. It is also used when only ranking values can be given. The ranks are obtained by allocating the integers 1 to n in place of the values x_i. For the smallest value of the x_i's we substitute 1, for the next smallest 2, for the next smallest 3, ... for the largest n (or vice versa in descending order). Similarly, for the values y_i, in ascending or descending order in conformity with the treatment of the x_i's. Suppose that the pair of values (x_i, y_i) have been given the ranks (q, s). Since q and s both take the integral values 1 to n:

$$\sum q = \sum s = \frac{n(n+1)}{2}, \quad \sum q^2 = \sum s^2 = \frac{n(n+1)(2n+1)}{6}$$

Also $\sum(q - s)^2 = \sum q^2 - 2\sum qs + \sum s^2$

i.e. $\sum qs \quad = \frac{1}{2}\sum q^2 + \frac{1}{2}\sum s^2 - \frac{1}{2}\sum(q - s)^2$

Thus we can write the regression coefficient

$$r = \frac{n\sum qs - \sum q \cdot \sum s}{\sqrt{\{[n\sum q^2 - (\sum q)^2][n\sum s^2 - (\sum s)^2]\}}}$$

as $\qquad r_s = \frac{n\frac{1}{2}\sum q^2 + n\frac{1}{2}\sum s^2 - \sum q \sum s - \frac{1}{2}n\sum(q - s)^2}{\sqrt{\{[n\sum q^2 - (\sum q)^2][n\sum s^2 - (\sum s)^2]\}}}$

but $\sum q = \sum s$, and $\sum q^2 = \sum s^2$

thus $r_s = \dfrac{n \sum q^2 - (\sum q)^2 - \dfrac{1}{2} n \sum (q - s)^2}{n \sum q^2 - (\sum q)^2}$

$$= 1 - \dfrac{\dfrac{1}{2} n \sum (q - s)^2}{n \sum q^2 - (\sum q)^2}$$

$$= 1 - \dfrac{\dfrac{1}{2} n \sum (q - s)^2}{\dfrac{n^2(n + 1)(2n + 1)}{6} - \dfrac{n^2(n + 1)^2}{4}}$$

$$= 1 - \dfrac{6 \sum D^2}{n(n^2 - 1)} \text{ where } D = (q - s)$$

This is *Spearman's rank correlation coefficient*. The exact distribution of r_s has been tabulated and Kendall has given tables of the frequency function $\sum (q - s)^2$ (the random component of r_s) for $n = 4(1)\ 10$ (the 'tail' entries in Kendall's tables are reproduced in the Biometrika tables). Beyond $n = 10$ the same test as that used for the product moment correlation coefficient r can be used.

Perfect positive correlation is obtained when $q = s$ for all pairs of values and then $r_s = 1$.

Perfect negative correlation is obtained when $q = n - s + 1$ for all pairs of values and then $r_s = -1$.

When the correlation is not perfect r_s lies between $+1$ and -1, i.e. $-1 \leqslant r_s \leqslant +1$.

Example 13.6. Find the rank correlation coefficient between the number of vehicles and the number of casualties in *Table 13.6*, and test the null hypothesis $\rho_s = 0$.

$$\sum (q - s)^2 = 8$$

$$r_s = 1 - \dfrac{6 \times 8}{10 \cdot 99}$$

$$= 0 \cdot 966$$

Using the product moment correlation test [see equation (13.10)], we form

$$t = \sqrt{\left[\dfrac{r^2(n - 2)}{1 - r^2} \right]} = \sqrt{\left(\dfrac{0 \cdot 966^2 \times 8}{1 - 0 \cdot 966^2} \right)} = \sqrt{\dfrac{7 \cdot 4648}{0 \cdot 0669}}$$

$$= 10 \cdot 57$$

Table 13.6

No. of vehicles x	q	No. of casualties y	s	$|q - s|$
2·6	10	138	10	0
3·1	9	163	8	1
3·5	8	166	7	1
3·7	7	153	9	2
4·1	6	177	6	0
4·4	5	201	5	0
4·9	3	208	4	1
5·8	1	238	1	0
5·3	2	226	2	0
4·6	4	216	3	1

Since $t_{1\%}(8) = 3\cdot36$, and $10\cdot57 > 3\cdot36$ we can reject, at the 1 per cent level, the null hypothesis that there is no correlation.

Using Kendall's tables we only require to find $\sum(q - s)^2$ and we find that for $n = 10$ then $\Pr[(q - s)^2 < 8]$ is less than $0\cdot001$. For a two-tailed test this represents a $0\cdot2$ per cent level of significance and we have the same result as before.

Comparing the two tests for $n = 10$ we find that Kendall's tables give a 1 per cent two-tailed critical value for r_s as $0\cdot794$, while the 't' test gives a critical value of $0\cdot765$.

Example 13.7. Is there any correlation between two judges X and Y who have placed contestants in a competition in the orders given in *Table 13.7*?

Table 13.7

Contestant	A	B	C	D	E	F
Judge X	2	1	4	3	6	5
Judge Y	1	2	6	3	5	4
$|X - Y|$	1	1	2	0	1	1

$$\sum(X - Y)^2 = 8$$

For $n = 6$ Kendall's tables give $\Pr[\sum(X - Y)^2 < 8]$ is equal to 0·051. This represents a 10 per cent level of significance for a two-tailed test and we cannot reject the null hypothesis of zero correlation.

13.6. RANKING OF EQUAL VARIATES

If there are two or more variates with the same value give each variate the average of the ranks they would have taken if they had been slightly different. For example see *Table 13.8*.

Table 13.8

	10·4	10·3	10·1	10·3	10·2
Rank	5	3·5	1	3·5	2

This will mean that the value of r_s is now approximate since the formula for $\sum i^2$ does not now apply.

13.7. KENDALL'S COEFFICIENT OF RANK CORRELATION (r_k)

An alternative coeffcient of rank correlation is obtained if we consider that the ranks of y are arrayed in natural order $1, 2, \ldots n$. The associated ranks of x, $x_1, x_2, \ldots x_n$ are a permutation of $1, 2, \ldots n$ and we can measure the disarray by counting the number of inversions of order among them*. The number of such inversions Q may range from 0 to $\frac{1}{2}n(n - 1)$ these limits being reached if the order of the x's is $1, 2, \ldots n$ or $n, n - 1, \ldots 2, 1$. Kendall's coefficient is defined as:

$$r_k = 1 - \frac{4Q}{n(n - 1)}$$

The distribution of r_k has mean zero and variance $\dfrac{2(2n + 5)}{9n(n - 1)}$ and it tends rapidly to normality. Kendall gives the exact distribution function for $n = 4(1)10$. Above $n = 10$ the normal approximation is adequate.

Example 13.8. Consider Example 13.7; if we order the ranks of judge Y we have *Table 13.9*.

---

* Only adjacent numbers may be interchanged.

Table 13.9

Judge Y	1	2	3	4	5	6
Judge X	2	1	3	5	6	4

We put Judge X's ranks in order by changing two adjacent ranks at a time, i.e.

$$\begin{array}{cccccc} 1 & 2 & 3 & 5 & 6 & 4 \\ 1 & 2 & 3 & 5 & 4 & 6 \\ 1 & 2 & 3 & 4 & 5 & 6 \end{array}$$

This has taken three inversions so $Q = 3$.

A simple way to find the number of inversions required is to arrange the Y ranks in natural order and put the corresponding X ranks underneath. Join corresponding numbers by straight lines and count the number of intersections of the lines to obtain Q, viz.:

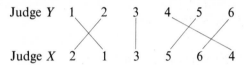

There are three intersections so $Q = 3$.

$$\therefore \qquad r_k = 1 - \frac{4 \times 3}{6 \times 5} = 0.6$$

Kendall's tables give $\Pr(Q \leqslant 3)$ as 0.068 so that the test is not significant at the 5 per cent level.

Exercise 13(a)

1. Calculate Spearman's and Kendall's rank correlation coefficients for the data given in *Table 13.10*.

Table 13.10

X	5	6	2	8	7	9	1	3	10	4	11	12
Y	4	6	7	3	1	2	10	12	5	9	8	11

Test both coefficients using the appropriate approximate tests. Use the null hypothesis $\rho = 0$.

EXERCISES 13

1. Show that for a bivariate distribution of n pairs (x_r, y_r) $r = 1, 2, \ldots n$

(a) $\bar{x} \sum y_r = \bar{y} \sum x_r$

(b) $\sum (x_r - \bar{x})(y_r - \bar{y}) = \sum x_r y_r - \dfrac{(\sum x_r)(\sum y_r)}{n}$

(c) $\sum (x_r + y_r)^2 = (n - 1)(s_x^2 + 2s_{x,y} + s_y^2) + n(\bar{x} + \bar{y})^2$

2. Find the linear equation connecting the pairs of x and y (take x as the independent variable) in *Table 13.11*.

Table 13.11

x	1	2	3	4	5
y	2	5	9	13	14

3. Find the lines of regression of x on y and y on x for the 12 pairs of values given in *Table 13.12*.

Table 13.12

x	84	82	82	85	89	90	88	97	83	89	98	99
y	78	77	85	88	87	82	81	77	76	83	97	93

4. To find the resistance of a wire it was connected in a Wheatstone bridge circuit and the readings obtained are given in *Table 13.13*.

Table 13.13

x	0·1	0·2	0·3	0·4	0·5	0·6	0·7	0·8	0·9	1·0
y	0·270	0·476	0·202	0·910	1·106	1·350	1·560	1·742	1·994	2·182

Assuming that y and x satisfy an equation of the form $y = rx + R$ where R is the resistance of the rest of the circuit and r is the resistance of the wire, find the values of r and R.

5. Show that the acute angle between the two regression lines x on y and y on x is $\tan^{-1} \left[\dfrac{(1 - r^2) s_x s_y}{r(s_x^2 + s_y^2)} \right]$.

6. Find the minimum value of r which would, when tested against the null hypothesis $\rho = 0$, give a significant result at the 5 per cent level of significance given samples of size 10, 20, 50 and 100.

7. From a given set of data it is found that the regression line of y on x is $y = 1\cdot2\ x$ and that for x on y is $x = 0\cdot6\ y$. Calculate the correlation coefficient r and the value of s_x/s_y.

8. A random sample of 12 pairs of observations from a normal population gives a correlation coefficient of $0\cdot35$. Is this a significant result when testing (a) the null hypothesis $\rho = 0$, (b) the null hypothesis $\rho = 0\cdot8$?

9. What is the least value of r in a sample of 22 pairs that gives ρ significantly different from zero at the 5 per cent level?

10. Find the regression equations of y on x and x on y for the bivariate distribution given in *Table 13.14*.

Table 13.14

		5	10	15	20	25	30	35	40
	5	1	1	—	—	—	—	—	—
	10	—	3	2	1	—	—	—	—
y	15	1	1	4	4	2	1	—	—
	20	—	—	1	2	5	8	8	2
	25	—	—	—	—	1	4	7	1

with x across the top.

11. Calculate the correlation coefficient for the data given in *Table 13.15*. Also test the null hypothesis that $\rho = 0\cdot7$.

12. Let X denote the order of merit of 10 students in one examination and Y the order of merit in a second examination, the corresponding values of X and Y being given in *Table 13.16*.

Find Spearman's rank correlation coefficient and Kendall's rank correlation coefficient.

13. The orders of merit of 20 competitors given by two judges are given in *Table 13.17*.

Test the null hypothesis that there is no correlation between the judges using Spearman's rank correlation coefficient.

14. Find a rank correlation coefficient for the paired data of question 3.

15. By using random number tables arrange the 15 integers 1 to 15 in random order. Repeat this process and obtain a second random order. By pairing the numbers in the order in which they were put obtain Spearman's rank correlation coefficient and test the hypothesis

Table 13.15

Height x

Weight y / Centre of interval	61	62	63	64	65	66	67	68	69	70	71	72
110	—	—	1	1	—	—	—	—	—	—	—	—
120	—	1	1	11	8	1	—	—	—	—	—	—
130	2	10	18	32	25	20	4	—	—	—	—	—
140	—	1	20	27	50	44	12	3	—	—	—	—
150	—	—	2	8	30	83	55	12	2	—	—	—
160	—	—	—	2	31	61	84	32	4	1	—	—
170	—	—	—	—	—	11	31	66	18	4	—	—
180	—	—	—	—	—	—	8	44	33	10	3	—
190	—	—	—	—	—	—	—	1	27	22	3	—
200	—	—	—	—	—	—	—	—	1	8	8	1
210	—	—	—	—	—	—	—	—	—	—	1	1

Table 13.16

X	1	2	3	4	5	6	7	8	9	10
Y	2	1	3	5	4	6	10	7	9	8

Table 13.17

Judge A	19	20	5	15	7	18	3	12	10	4
Judge B	12	9	20	16	13	5	10	2	4	18

Judge A	17	2	8	6	1	11	14	9	13	16
Judge B	7	8	14	11	19	17	1	6	15	3

of zero correlation. Explain why the test could give a significant result in this case.

16. *Table 13.18* gives the initial weights (in 0·001 g) of larvae of a wood-boring beetle, and their weights after feeding on slightly decayed wood for 35 days. Determine the regression coefficient of final on initial weight, and test whether it differs significantly from 1. State your conclusions regarding the nutritive value of slightly decayed wood.

Table 13.18

Larvae	1	2	3	4	5	6	7	8	9
Initial weight	74	77	79	84	95	98	112	120	128
Final weight	85	91	95	104	110	107	134	146	169

(Liv. U.)

17. Show how the product-moment correlation of two sets of rankings X_i and Y_i of n objects simplifies to the formula

$$r = 1 - 6 \sum_{i=1}^{n} [d_i^2/n(n^2 - 1)] \text{ where } d_i = X_i - Y_i$$

You may use the fact that the sum and the sum of squares of the first n natural numbers are $n(n + 1)/2$ and $n(n + 1)(2n + 1)/6$.

Two judges in a beauty contest ranked the different entries as shown in *Table 13.19*.

Table 13.19

Entry	A	B	C	D	E	F	G	H	I	J	K	L
Judge 1	4	1	2	3	6	7	5	8	9	12	10	11
Judge 2	7	2	1	5	4	8	3	6	9	10	11	12

Test whether the two sets of rankings are correlated. (Liv. U.)

18. Associated Television Limited: *Table 13.20* shows the charges in £ for 15-sec advertisements.

This shows how the charges for advertising, £y, made by ATV were related to the number of homes viewing, x, during the autumn of 1956. Thus the number 17 in the table indicates that there were 17 cases in which the charge was between £200 and £250 when between 200,000 and 300,000 homes were viewing.

Table 13.20

Charge (£y) for 15 sec *advertisement*	No. (x) of homes viewing ATV programmes (1000's)							
	100–	200–	300–	400–	500–	600–	700–	800 and over
0–	6							
50–	2	5	2					
100–	5	7	3	1	1			
150–	2	6	1	3	1	2		
200–	1	17	7	0	1	0		
250–				2	1	1		
300–					1	1		
350–					1	0		
400–						3	1	
450 and over								

For each range of x calculate to the nearest integer the mean value of y. Plot your results on a scatter diagram and sketch in the line of regression of y on x.

Estimate the mean charge made by ATV for a 15-sec advertisement at a time when 500,000 homes were viewing. (J.M.B.)

19. Selected vehicle-manufacturing industries: estimated number of employees in Great Britain (thousands) are given in *Table 13.21*.

Table 13.21

	Manufacture of motor vehicles and cycles x	Manufacture of parts and accessories for motor vehicles and aircraft y
1948 June	279	92
1949 June	292	95
1950 June	297	115
1951 June	299	124
1952 June	299	139
1953 June	295	143
1954 June	311	157
1955 June	329	173
1956 June	321	173

Use the above table to plot a scatter diagram of values of y against values of x. On a diagram mark the point $(\bar{x}, \bar{y})$ whose coordinates are the means of the values of x, y respectively. Draw through the point $(\bar{x}, \bar{y})$, as accurately as you can by eye the line of regression of y on x.

Determine the gradient of the line, and comment on its meaning.

(J.M.B.)

20. Five groups of locusts, each containing 120, were exposed to a lethal spray in various concentrations. The deaths resulting are given in *Table 13.22*.

Table 13.22

Concentration (multiple of standard)	1·2	1·4	1·6	1·8	2·0
Number of deaths	38	52	46	67	66

For each concentration find the percentage of locusts dying. Plot this percentage against the concentration on a scatter diagram and sketch in the line of regression, taking concentration as your independent variable.

Estimate the percentage of locusts which might be expected to die at a concentration of 1·5. Assuming that your estimate is in fact correct, give limits between which the number of locusts dying in a sample of 120 might be expected to vary in 90 per cent of trials at this concentration.   (J.M.B.)

21. *Table 13.23* shows the Intelligence Quotient (I.Q.) and the mark obtained in an examination for each of 10 children. Evaluate the product-moment correlation coefficient between the I.Q.s and the examination marks.

Table 13.23

I.Q.	145	100	100	112	138	133	123	116	127	106
Mark	75	35	52	36	80	90	71	54	52	55

(The mean I.Q. is 120 and the mean mark 60.)   (J.M.B.)

14

MATHEMATICAL EXPECTATION, VARIANCE AND COVARIANCE

14.1. INTRODUCTION

IN this chapter the idea of mathematical expectation and its relationship with mean, variance and covariance, first introduced in sections 3.6, 6.4, 7.3 and 13.1, is considered in greater depth. Also several important results, which have only been quoted earlier, are proved.

14.2. VARIANCE

The variance of x [written var(x)], where x is a member of a population was defined as the second moment about the mean. Thus if $h(x)$ is the continuous probability density function associated with the population and $\mathscr{E}(x) = \mu$ then

$$\text{var}(x) = \int_a^b (x - \mu)^2 \, h(x) \, dx$$

If the population is discrete with probability p_i associated with x_i ($i = 1, 2, \ldots n$) then

$$\text{var}(x) = \sum_{i=1}^n p_i(x_i - \mu)^2$$

By the definition of expectation

$$\int_a^b (x - \mu)^2 \, h(x) \, dx = \mathscr{E}(x - \mu)^2$$

$$\therefore \qquad \text{var}(x) = \mathscr{E}(x - \mu)^2$$
$$= \mathscr{E}(x^2 - 2x\mu + \mu^2)$$
$$= \mathscr{E}(x^2) - 2\mu\mathscr{E}(x) + \mu^2$$
$$= \mathscr{E}(x^2) - \mu^2 \qquad \qquad \ldots . (14.1)$$
$$= \mathscr{E}(x^2) - [\mathscr{E}(x)]^2$$

Also since $\mathscr{E}(ax) = a\mathscr{E}(x)$ (from the definition of expectation)
$$= a\mu \ (a \text{ constant})$$
$$\text{var}(ax) = \mathscr{E}(ax - a\mu)^2 = a^2\mathscr{E}(x - \mu)^2$$
$$= a^2 \, \text{var}(x) \qquad \qquad \ldots . (14.2)$$

14.3. COVARIANCE

If x_1 and x_2 are two variables and $\mathscr{E}(x_1) = \mu_1$, and $\mathscr{E}(x_2) = \mu_2$ then the *covariance* of x_1 and x_2 [written $\text{cov}(x_1, x_2)$] is defined by

$$
\begin{aligned}
\text{cov}(x_1, x_2) &= \mathscr{E}(x_1 - \mu_1)(x_2 - \mu_2) \\
&= \mathscr{E}(x_1 x_2) - \mathscr{E}(\mu_1 x_2) - \mathscr{E}(x_1 \mu_2) + \mathscr{E}(\mu_1 \mu_2) \\
&= \mathscr{E}(x_1 x_2) - \mu_1 \mathscr{E}(x_2) - \mu_2 \mathscr{E}(x_1) + \mu_1 \mu_2
\end{aligned}
$$

substituting expectations for μ_1 and μ_2 we have

$$
\text{cov}(x_1, x_2) = \mathscr{E}(x_1 x_2) - \mathscr{E}(x_1)\,\mathscr{E}(x_2)
$$

If x_1 and x_2 are statistically independent then

$$
\text{cov}(x_1, x_2) = 0
$$

Note: converse is only true if x_1 and x_2 are normally distributed.

14.4. EXPECTATION AND VARIANCE OF THE SUM AND DIFFERENCE OF TWO VARIATES

Let x be a member of a population whose mean is μ_x and variance σ_x^2 and let y be a member of a population whose mean is μ_y and variance σ_y^2.

$$
\begin{aligned}
\mathscr{E}(x + y) &= \mathscr{E}(x) + \mathscr{E}(y) = \mu_x + \mu_y \\
\mathscr{E}(x - y) &= \mathscr{E}(x) - \mathscr{E}(y) = \mu_x - \mu_y \\
\text{var}(x + y) &= \mathscr{E}[(x + y - \mu_x - \mu_y)^2] \\
&= \mathscr{E}[(x - \mu_x)^2] + \mathscr{E}[(y - \mu_y)^2] + 2\mathscr{E}[(x - \mu_x)(y - \mu_y)] \\
&= \text{var}(x) + \text{var}(y) + 2\,\text{cov}(x, y) \\
\text{var}(x - y) &= \mathscr{E}[(x - y - \mu_x + \mu_y)^2] \\
&= \mathscr{E}[(x - \mu_x)^2] + \mathscr{E}[(y - \mu_y)^2] - 2\mathscr{E}[(x - \mu_x)(y - \mu_y)] \\
&= \text{var}(x) + \text{var}(y) - 2\,\text{cov}(x, y)
\end{aligned}
$$

If x and y are independent $\text{cov}(x, y) = 0$, and

$$
\text{var}(x + y) = \text{var}(x - y) = \text{var}(x) + \text{var}(y) \qquad \dots (14.3)
$$

As a corollary to this, if $x_1, x_2, \dots x_n$ are n independent variables

$$
\text{var}(\Sigma x_i) = \Sigma[\text{var}(x_i)] \qquad \dots (14.4)
$$

Exercises 14a

1. Show that $\text{var}(ax + by) = a^2\text{var}(x) + b^2(\text{var}(y) + 2ab\text{cov}(x, y)$.
2. Show that $\text{var}(x + y - z) = \text{var}(x) + \text{var}(y) + \text{var}(z) + 2\text{cov}(x, y) - 2\text{cov}(x, z) - 2\text{cov}(y, z)$.

3. Show that by the definition of covariance $\text{cov}(x, x) = \text{var}(x)$; hence using the relationship $\text{var}(x + y) = \text{var}(x) + \text{var}(y) + 2\text{cov}(x, y)$ deduce that $\text{var}(2x) = 2^2\text{var}(x)$ [this verifies equation (14.2)].

Example 14.1. If $x_1, x_2, \ldots x_n$ are a random sample from a population with mean $\mu = \mathscr{E}(x)$ and variance σ^2 show that $\text{var}(\bar{x}) = \sigma^2/n$ and $s^2 = \Sigma(x - \bar{x})^2/(n - 1)$ is an unbiased estimate of σ^2.

$$\text{var}(\bar{x}) = \text{var}\left(\frac{\Sigma x}{n}\right) = \frac{1}{n^2}\text{var}(\Sigma x) \text{ [see equation (14.2)]}$$

$$= \frac{1}{n^2}\Sigma\,\text{var}(x) \text{ [see equation (14.4)]}$$

$$= \frac{n}{n^2}\text{var}(x)$$

$$= \frac{\sigma^2}{n} \qquad\qquad \ldots(14.5)$$

Note: since $\mathscr{E}(\bar{x}) = \mu$ we also have that:

$$\text{var}(\bar{x}) = \mathscr{E}(\bar{x} - \mu)^2\,\mathscr{E}(\bar{x}^2) - [\mathscr{E}(\bar{x})]^2$$
$$= \mathscr{E}(\bar{x}^2) - \mu^2 \qquad\qquad \ldots(14.6)$$

$$\mathscr{E}(s^2) = \frac{1}{(n-1)}\mathscr{E}[\Sigma(x - \bar{x})^2] = \frac{1}{(n-1)}\mathscr{E}[\Sigma(x^2) - n\bar{x}^2]$$

$$= \frac{1}{(n-1)}[\Sigma\mathscr{E}(x^2) - n\mathscr{E}(\bar{x}^2)]$$

$$= \frac{1}{(n-1)}[n\mathscr{E}(x^2) - n\mathscr{E}(\bar{x}^2)]$$

$$= \frac{n}{(n-1)}\{\mathscr{E}(x^2) - \mu^2 - [\mathscr{E}(\bar{x}^2) - \mu^2]\}$$

$$= \frac{n}{(n-1)}[\text{var}(x) - \text{var}(\bar{x})] \text{ [see equations (14.1) and (14.6)]}$$

$$= \frac{n}{(n-1)}\left[\sigma^2 - \frac{\sigma^2}{n}\right] \text{ [see equation (14.5)]}$$

$$= \sigma^2 \qquad\qquad \ldots(14.7)$$

From the preceding example it is seen that if samples of size n are taken from a population mean μ and variance σ^2, then the population made up of all the means from all possible samples size n is distributed with mean μ and variance σ^2/n.

I

Example 14.2. Verify the results of equations (3.22), (14.5), (14.7) and question 5 Exercises 14(b). Suppose that we have an infinite population containing equal numbers of 1, 3, 5, 7 and take samples of size 2.

The population mean $\mu = 16/4 = 4$ and the population variance is

$$\sigma^2 = (84 - 64)/4 = 5.$$

All possible samples of size 2 are: 1,1 1,3 1,5 1,7 3,3 3,5 3,7 5,5 5,7 7,7 3,1 5,1 7,1 5,3 7,3 7,5.
The means of these samples are respectively:

1, 2, 3, 4, 3, 4, 5, 5, 6, 7, 2, 3, 4, 4, 5, 6.

The mean and variance of this distribution of means are

Mean $= \dfrac{64}{16} = 4 = \mu$ [this verifies equation (3.22)]

Variance $= \dfrac{1}{16}(296 - 64 \times 4) = \dfrac{40}{16} = \dfrac{5}{2}$

$= \dfrac{\text{population variance}}{\text{sample size}} = \dfrac{\sigma^2}{n}$ [this verifies equation (14.5)]

The histogram for the distribution of means is obtained by plotting the frequencies given in *Table 14.1*:

Table 14.1

$\bar{x}$	1	2	3	4	5	6	7
f	1	2	3	4	3	2	1

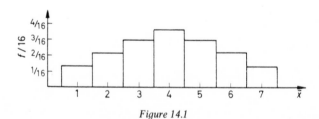

Figure 14.1

The histogram approximates very well to a normal distribution. The variances of the samples are given by $\Sigma(x - \bar{x})^2/(n - 1)$ and are respectively: 0, 2, 8, 18, 0, 2, 8, 0, 2, 0, 2, 8, 18, 2, 8, 2; the mean of these sample variances is $80/16 = 5$ [this verifies equation (14.7)]. Calculating $s^2 = \Sigma(x - \mu)^2/n$ for each sample we obtain: 9, 5, 5, 9,

1, 1, 5, 1, 5, 9, 5, 5, 9, 1, 5, 5. The mean of these variances is $80/16 = 5$ [this verifies the result quoted in question 5 Exercises 14(b)].

Exercises 14b

1. In Example 14.2 take samples of size 3 and show that the same conclusions are reached.

2. Consider an infinite population containing equal numbers of any five different integers. Take samples of size 2 and show that the above facts are again true.

3. (a) Use random numbers to obtain 100 random samples from the normal distribution $N(0,1)$.

(b) Plot a histogram of the sample means and find the mean and variance of this sample of means. Compare your results with the expected values.

4. Repeat question 3 using the rectangular distribution $h(x) = 1$ $(0 \leqslant x \leqslant 1)$ instead of the normal distribution.

5. If $x_1, x_2, \ldots x_n$ are a random sample from a population mean μ, variance σ^2, show that $s^2 = \Sigma(x - \mu)^2/n$ is an unbiased estimate of σ^2.

Example 14.3. Test the formula for the expectation and variance of the sum and difference of two variates by considering a population made up of 1's and 3's (x) and a population made up of 5's and 7's (y). Consider the two cases (a) x and y independent, (b) x and y related by the equation $y = x + 4$.

(a) If x and y are independent all possible values of $(x + y)$ are: $(1 + 5), (1 + 7), (3 + 5), (3 + 7)$, that is 6, 8, 8, 10.

$$\mathscr{E}(x) = \frac{1 + 3}{2} = 2; \quad \mathscr{E}(y) = \frac{5 + 7}{2} = 6$$

$$\mathscr{E}(x + y) = \frac{6 + 8 + 8 + 10}{4} = 8$$

$\therefore$ (for this case) $\mathscr{E}(x + y) = \mathscr{E}(x) + \mathscr{E}(y)$

Furthermore, var$(x) = 1$ var$(y) = 1$

$$\text{var}(x + y) = \tfrac{1}{4}(8) = 2 = \text{var}(x) + \text{var}(y)$$

If x and y are independent, all possible values of $(x - y)$ are $(1 - 5)$, $(1 - 7), (3 - 5), (3 - 7)$, that is $-4, -6, -2, -4$.

$$\mathscr{E}(x) = 2 \text{ and } \mathscr{E}(y) = 6 \text{ as before}$$

$$\mathscr{E}(x - y) = -16/4 = -4 = \mathscr{E}(x) - \mathscr{E}(y)$$

$$\text{var}(x - y) = \tfrac{1}{4}.8 = 2 = \text{var}(x) + \text{var}(y)$$

(b) In this case since $y = x + 4$ when $x = 1$, $y = 5$, and when

$x = 3, y = 7$, and hence the only possible values of $(x + y)$ are $(1 + 5)$ and $(3 + 7)$, that is 6 and 10.

$$\text{cov}(x, y) = \mathscr{E}(xy) - \mathscr{E}(x)\,\mathscr{E}(y) = \frac{1 \times 5 + 3 \times 7}{2} - 2 \times 6 = 1$$

$$\mathscr{E}(x + y) = 16/2 = 8 = \mathscr{E}(x) + \mathscr{E}(y)$$
$$\text{var}(x + y) = \tfrac{1}{2}\,.\,8 = 4 = \text{var}(x) + \text{var}(y) + 2\,\text{cov}(x, y)$$

The only possible values of $(x - y)$ are $(1 - 5), (3 - 7)$ that is $-4, -4$.

$$\mathscr{E}(x - y) = -4 = \mathscr{E}(x) - \mathscr{E}(y)$$
$$\text{var}(x - y) = 0 = \text{var}(x) + \text{var}(y) - 2\,\text{cov}(x, y)$$

Exercises 14c

1. Using the two populations $-2, 0, 2$ (x) and $0, 4$ (y), illustrate the theory as shown above in Example 14.2 taking first x and y as independent and then connected by the equation $y = x^2$.

2. Given three independent populations $0, 4$ (x), $1, 3$ (y) $0, 3$ (z) show that:
(a) $\mathscr{E}(x + y - z) = \mathscr{E}(x) + \mathscr{E}(y) - \mathscr{E}(z)$
(b) $\text{var}(x + y - z) = \text{var}(x) + \text{var}(y) + \text{var}(z)$.

Example 14.4. Rectangular sheets of metal of length L and width W are being mass-produced. The mean length is 4 in with a standard deviation of 0·4 in, while the mean width is 2 in with a standard deviation of 0·3 in. Find the mean and standard deviation of the perimeter if the length and width can be assumed independent.

$$p = 2(L + W)$$
$$\mathscr{E}(p) = 2\,[\mathscr{E}(L) + \mathscr{E}(W)] = 2(4 + 2) = 12\,\text{in}$$
$$\text{var}(p) = 2^2\,[\text{var}(L) + \text{var}(W)] = 4(0\cdot16 + 0\cdot09) = 1$$

The standard deviation of p is 1 in.

Example 14.5. Find the mean and standard deviation of the area of the metal sheets described in Example 14.4.

$$A = L \times W$$
$$\mathscr{E}(A) = \mathscr{E}(L) \times \mathscr{E}(W) = 4 \times 2 = 8$$
$$\text{var}(A) = \text{var}(LW) = \mathscr{E}(L^2 W^2) - [\mathscr{E}(LW)]^2$$
$$= \mathscr{E}(L^2)\,.\,\mathscr{E}(W^2) - [\mathscr{E}(L)]^2\,.\,[\mathscr{E}(W)]^2$$
$$= \mathscr{E}(L^2)\,.\,\mathscr{E}(W^2) - 16\,.\,4$$

Now $\text{var}(L) = \mathscr{E}(L^2) - [\mathscr{E}(L)]^2 \therefore \mathscr{E}(L^2) = \text{var}(L) + [\mathscr{E}(L)]^2$

i.e. $\mathscr{E}(L^2) = 0{\cdot}16 + 16{\cdot}0 = 16{\cdot}16$

similarly, $\mathscr{E}(W^2) = \text{var}(W) + [\mathscr{E}(W)]^2 = 0{\cdot}09 + 4 = 4{\cdot}09$

$$\therefore \quad \text{var}(A) = 16{\cdot}16 \times 4{\cdot}09 - 16 \times 4$$
$$= 2{\cdot}0944$$

$\therefore$ The standard deviation of $A = 1{\cdot}45$

Example 14.6. An automatic canning machine delivers sealed cans of meat whose total weight is normally distributed with a mean value of 19·80 oz and standard deviation 0·12 oz. The weight of the empty cans (plus lids) delivered to the machine is normally distributed with a mean value of 3·50 oz and standard deviation 0·05 oz. Show that approximately 1 per cent of the cans have contents whose weight is less than 16 oz.

Referring to *Figure 14.2* it is seen that $y = z - x$ and since z and x are independent we have that $\text{cov}(z, x) = 0$ and

$$\mathscr{E}(y) = \mathscr{E}(z) - \mathscr{E}(x) = 19{\cdot}80 - 3{\cdot}50 = 16{\cdot}30 \text{ oz}$$

$$\text{also } \text{var}(y) = \text{var}(z) + \text{var}(x) = 0{\cdot}0144 + 0{\cdot}0025$$
$$= 0{\cdot}0169$$

and standard deviation of $y = 0{\cdot}13$ oz.

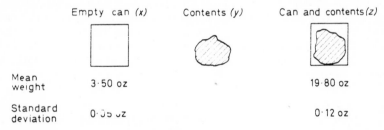

	Empty can (x)	Contents (y)	Can and contents (z)
Mean weight	3·50 oz		19·80 oz
Standard deviation	0·05 oz		0·12 oz

Figure 14.2

Since the two variables z and x are both normally distributed it follows that y is normally distributed with mean 16·30 oz and standard deviation 0·13 oz.

We require $\text{Pr}(y < 16)$ and therefore form the standardized variate

[see section 8.2(g)] $\quad y_1 = \dfrac{16 - 16{\cdot}30}{0{\cdot}13} = -2{\cdot}3077$

Now $\Pr(y_1 < -2.3077) = \Pr(y_1 > 2.3077) = 1 - 0.9895$

$$= 0.0105$$

$$\simeq 1 \text{ per cent}$$

EXERCISES 14

1. The density of pieces of metal being mass-produced has a mean of 3.4 g/cm^3 and a standard deviation of 0.2. The volumes of the pieces have a mean of 25 cm^3 and a standard deviation of 0.15. Find the mean and standard deviation of the mass of the pieces.

2. In Example 14.6 explain the fallacy in arguing that:

$$\text{`}z = x + y \therefore \text{var}(z) = \text{var}(x) + \text{var}(y)$$

$$\therefore \text{var}(y) = \text{var}(z) - \text{var}(x)\text{'}$$

3. The lengths of a large number of planks of timber are normally distributed with a mean length of 18.00 ft and a standard deviation of 0.30 ft. Four identical smaller pieces (nominal length 4 ft) are cut from each plank. These smaller pieces are measured and found to have a mean length of 4.10 ft and standard deviation 0.10 ft, and to be approximately normally distributed. Find the mean length of the scrap wood, the standard deviation of this mean, and the probability that the length of a piece of scrap wood is greater than 2.0 ft.

4. Four swimmers form a relay team for an eight lengths relay race. During training their mean times for a length are respectively 15.4 sec, 15.8 sec, 15.6 sec and 15.9 sec. The corresponding standard deviations are each 0.4 sec. Assuming that the individual times are normally distributed and independent of each other estimate the mean and standard deviation of the times in which they swim eight lengths if each one swims two identically timed lengths. Show that they have a 1 in 4 chance of beating a record of 2 min 4.32 sec for the distance.

5. A canning machine puts a nominal 8 oz of vegetables into a can and then seals the can. The weight of the empty cans (plus lids) and the weight of the contents are both normally distributed with actual means 2.60 oz and 8.04 oz and standard deviations 0.06 oz and 0.08 oz respectively. Assuming that no weight is lost during the sealing process, find the probability that a full can weighs less than 10.50 oz.

6. A composite resistance of 2500 Ω is made up from two identical 1000 Ω and five identical 100 Ω resistances. From tests it is known that the subsidiary resistances actually have mean values of 1001 Ω and 99 Ω with standard deviations of 24 Ω and 10 Ω respectively. Assuming normal distributions, find the probability that the composite resistance will have a value between 2500 ± 50 Ω.

7. Find the mean and variance of the population given in *Table 14.2*, and draw a sample of 10 from it, using a table of random sampling numbers. Does the mean of your sample fall within the range you would reasonably expect?

Table 14.2

Variate value	6	7	8	9	10
Probability	0·05	0·15	0·35	0·25	0·20

(Liv. U.)

8. Denote by m the median of the distribution with probability density function $f(x)$; thus

$$\int_{-\infty}^{m} f(x)\, dx = \int_{m}^{\infty} f(x)\, dx = \frac{1}{2}$$

Prove that

$$\mathscr{E}\,|x - t| = \mathscr{E}\,|x - m| + 2\int_{m}^{t} (t - x) f(x)\, dx \text{ if } t > m$$
$$= \mathscr{E}\,|x - m| + 2\int_{t}^{m} (x - t) f(x)\, dx \text{ if } t < m$$

Deduce that the mean absolute deviation from t is a minimum when t is the median m. (Liv. U.)

9. Prove that

$$\mathscr{E}(xy) = \mathscr{E}(x)\,.\,\mathscr{E}(y) + \mathscr{E}\,\{[x - \mathscr{E}(x)][y - \mathscr{E}(y)]\}$$

The average wage per hour of a group of workers is three shillings, and the average number of hours worked per week is 40. In the light of the above relation, examine the accuracy of the deduction that the average weekly earnings of workers in the group is 120 shillings. (Liv. U.)

10. Three dice, each numbered in the usual way from one to six, are coloured white, red and blue respectively. After casting them a boy 'scores' in the following way. To the white number he adds twice the red number and then subtracts the blue number. Thus a white three, a red four and a blue two would score

$$3 + 8 - 2 = 9$$

Assuming that the boy casts the dice a large number of times, calculate the mean and variance of the scores. (J.M.B.)

11. Four athletes specialize in running 220 yd, 220 yd, 440 yd and 880 yd respectively. They train as a team for a 1 mile medley relay race. During training their mean times for their respective distances are 23·9 sec, 24·1 sec, 53·6 sec and 2 min 7·4 sec, and the corresponding

standard deviations are 0·3 sec, 0·3 sec, 0·8 sec and 1·8 sec. Assuming that their individual times are normally and independently distributed, estimate the mean and standard deviation of the times in which the team covers the mile. Also estimate the probability that, on any particular occasion, the time will be 3 min 45 sec or less.   (J.M.B.)

15

WEIGHTED AVERAGES, DEATH RATES AND TIME SERIES

15.1. WEIGHTED AVERAGES

THE arithmetic mean of n quantities $x_1, x_2, x_3, \ldots x_n$ is given by $\dfrac{x_1 + x_2 + \ldots x_n}{n}$. If each quantity is weighted by amounts $w_1, w_2, \ldots w_n$ then the *weighted arithmetic mean* is defined by

$$\frac{w_1 x_1 + w_2 x_2 + \ldots + w_n x_n}{w_1 + w_2 + \ldots + w_n}$$

or

$$\frac{\sum\limits_{r=1}^{n} w_r x_r}{\sum\limits_{r=1}^{n} w_r} \qquad \ldots (15.1)$$

The weights $w_1, w_2, w_3, \ldots w_n$ are used to give more (large value of w) or less (small value of w) importance to the effect of $x_1, x_2, \ldots x_n$ in calculating the final weighted average.

Example 15.1. A boy gained the following marks in his terminal examination: English 79, Mathematics 58, History 66, Geography 21, French 58, Art 62. For an assessment of his final standard it is agreed that the weighting of the subjects should be 4, 4, 3, 3, 2, 1 respectively. Calculate his weighted average.

From equation (15.1)

weighted average =

$$\frac{4 \times 79 + 4 \times 58 + 3 \times 66 + 3 \times 21 + 2 \times 58 + 1 \times 62}{4 + 4 + 3 + 3 + 2 + 1} = 58 \cdot 0$$

It will be noted that for a frequency distribution the average $\dfrac{\Sigma f x}{\Sigma f}$ is a weighted average of the values of x with the frequencies acting as the weights.

Although the weighted average usually used is the weighted arithmetic mean, a similar weighting can be applied to any mean, for

239

example the geometric mean (see section 5.5). If x_1, x_2 ... x_n have weights w_1, w_2, ... w_n then their weighted geometric mean is

$$\text{Weighted geometric mean} = \sqrt[N]{[(x_1)^{w_1} \times (x_2)^{w_2} \times \ldots \times (x_n)^{w_n}]}$$

$$\ldots (15.2)$$

$$\text{where } N = w_1 + w_2 + \ldots w_n$$

Taking logarithms we have that

$$\text{Log (W.G.M.)} = \frac{w_1 \log(x_1) + w_2 \log(x_2) + \ldots w_n \log(x_n)}{N}$$

$$= \frac{\sum_{r=1}^{n} w_r \log x_r}{\sum_{r=1}^{n} w_r} \qquad \ldots (15.3)$$

Or in words, the logarithm of the weighted geometrical mean of $x_1, x_2, \ldots x_n$ is the weighted arithmetic mean of their logarithms.

Example 15.2. Given the three numbers 1·09, 1·16, 1·75, with weights 3, 2, and 1 respectively find their weighted geometric mean.
From equation (15.3)

$$\text{Log (W.G.M.)} = \frac{3 \times \log 1·09 + 2 \times \log 1·16 + 1 \times \log 1·75}{3 + 2 + 1}$$

$$= \frac{3 \times 0·0374 + 2 \times 0·0645 + 0·2430}{6}$$

$$= 0·0807$$

therefore W.G.M. = 1·204.

Exercises 15a
1. Five lb of coffee are bought at 6s. 6d. per lb, 12 lb at 7s. 0d. per lb and 8 lb at 9s. 0d. per lb. Find the average price per lb.
2. The average marks in an examination in each of three classes are 58, 52, 41. There are respectively 43, 48 and 36 boys in each of the classes. Find the average examination mark in the three classes taken together.
3. A group of companies produce five different types of cars. They retail at £520, £650, £800, £900 and £1200 each respectively. In a certain period the group sells 5000, 4500, 3000, 2000 and 1000 respectively of each type. What is the average price of all the cars sold? (to the nearest £).

4. *Table 15.1* refers to the sales of coal from four coal depots for the months of September and October.

Table 15.1

	September Average price/ton (£'s)	No. tons sold (1000's)	October Average price/ton (£'s)	No. tons sold (1000's)
A	10·0	36	9·5	24
B	12·0	25	11·5	20
C	14·0	15	13·5	18
D	17·0	10	16·5	20

Note: the average price per ton for October was less than that for September in all four depots.

Calculate the average cost per ton in September and October for the total coal sold by the four depots. Comment on the result.

5. Find the weighted geometric mean of the four numbers 302, 217, 12·15, 0·073 using weights of 2, 3, 6, and 8 respectively.

15.2. INDEX NUMBERS

In many applications of statistics, especially in economics, we wish to compare different sets of data. To do this the data is reduced to purely relative numbers by comparing it with a fixed (base) value. Such relative numbers are the simplest index numbers, they are known as *price relatives*, and are generally expressed as percentages.

The selection of a base for the numbers is important and the purpose for which they are required often indicates the base to be used. The base need not refer to a single set of data but can be an average of several sets taken over a period of time, or of several sets from different locations, but it is important that it should fairly represent a normal or average situation which occurred not too long ago.

Example 15.3. Using 1948 as a base year calculate the price relatives for commodity A, from *Table 15.2* (the table contains more information than is required because it will be used later).

Price relatives for commodity A using 1948 as a base year are

$$1949 \qquad \frac{23 \cdot 0}{21 \cdot 0} \times 100 = 109 \cdot 5$$

$$1950 \qquad \frac{24 \cdot 2}{21 \cdot 0} \times 100 = 115 \cdot 2$$

Thus for commodity A price relatives are given in *Table 15.3*.

Table 15.2

Commodity	Price per stone to the nearest shilling			Weight
	1948	1949	1950	
A	21·0	23·0	24·2	7
B	5·2	7·1	6·8	5
C	103·7	120·9	97·4	1
D	85·6	84·3	86·1	3
Total	215·5	235·3	214·5	

Table 15.3

	1948	1949	1950
A	100	109·5	115·2

Similar price relatives can be obtained for B, C, and D and these are left as an exercise for the reader.

Another way of calculating a simple index number is to compare the total price of buying equal quantities of different types of goods with the total price some time later, or in a different region.

Example 15.4. Use the data in Example 15.3 to express the cost of buying one unit each of A, B, C, and D in 1949 and 1950 as an index number using 1948 as a base year.

The total cost of one unit each of A, B, C, and D is given in the table, therefore the required index numbers are as follows

$$1949 \quad \frac{235\cdot3}{215\cdot5} \times 100 = 109\cdot2$$

$$1950 \quad \frac{214\cdot5}{215\cdot5} \times 100 = 99\cdot5$$

The simple index numbers obtained in Example 15.4 do not take into account the relative importance of the different commodities. Different weights can be applied to the different commodities and a weighted mean found for each year, these can then be converted into indexes. This more complex index is often of more practical value than the simpler index, but it is often difficult to decide the weight to be applied to each commodity.

One of the best known indexes is the *cost of living index*, which, as its name implies, sets out to measure the cost of living. It is a very difficult matter, however, to decide which items are to be included and what weight is to be applied to each.

Example 15.5. From the data of Example 15.3, using weights 7, 5, 1, and 3 respectively for the commodities A, B, C and D, calculate a weighted mean for each of the three years 1948, 1949 and 1950. Express these weighted means as index numbers using 1948 as a base year.

(1948)

$$\text{Weighted mean} = \frac{7 \times 21 \cdot 0 + 5 \times 5 \cdot 2 + 1 \times 103 \cdot 7 + 3 \times 85 \cdot 6}{7 + 5 + 1 + 3}$$

$$= 33 \cdot 34 \qquad \qquad \dots \text{(i)}$$

(1949)

$$\text{Weighted mean} = \frac{7 \times 23 \cdot 0 + 5 \times 7 \cdot 1 + 1 \times 120 \cdot 9 + 3 \times 84 \cdot 3}{7 + 5 + 1 + 3}$$

$$= 35 \cdot 64 \qquad \qquad \dots \text{(ii)}$$

(1950)

$$\text{Weighted mean} = \frac{7 \times 24 \cdot 2 + 5 \times 6 \cdot 8 + 1 \times 97 \cdot 4 + 3 \times 86 \cdot 1}{7 + 5 + 1 + 3}$$

$$= 34 \cdot 94 \qquad \qquad \dots \text{(iii)}$$

Now, taking 1948 as the base year, the index numbers calculated from the above weighted means are

$$(1949) \frac{35 \cdot 64}{33 \cdot 34} \times 100 = 106 \cdot 9$$

$$(1950) \frac{34 \cdot 94}{33 \cdot 34} \times 100 = 104 \cdot 8$$

Sometimes instead of the actual cost the price relatives for the different commodities are known. In these cases the same result can be obtained by taking a weighted mean of the price relatives but using as weights numbers equal or proportional to the total expenditure on each commodity in the base year.

Example 15.6. Calculate index numbers for the years 1949 and 1950 using the price relatives obtained in Example 15.3 for the commodities A, B, C, and D. Use weights equal to the total expenditure on each item in 1948.

From Example 15.3 we have in *Table 15.4* the price relatives and the corresponding weights:

Table 15.4

	1948	1949	1950	*Weight*
A	100	109·5	115·2	$7 \times 21·0 = 147·0$
B	100	136·5	130·8	$5 \times 5·2 = 26·0$
C	100	116·6	93·9	$1 \times 103·7 = 103·7$
D	100	98·5	100·6	$3 \times 85·6 = 256·8$

(1949)

$$\text{Index No.} = \frac{109·5 \times 147 + 136·5 \times 26 + 116·6 \times 103·7 + 98·5 \times 256·8}{147 + 26 + 103·7 + 256·8}$$

$$= 106·9$$

(1950)

$$\text{Index No.} = \frac{115·2 \times 147 + 130·8 \times 26 + 93·9 \times 103·7 + 100·6 \times 256·8}{147 + 26 + 103·7 + 256·8}$$

$$= 104·8$$

which are the same as the answers in Example 15.5.

It can be proved in general that the two methods used in Examples 15.5 and 15.6 are equivalent.

Exercises 15b

1. *Table 15.5* shows the average export price of coal in each of the years 1905–1914.

Table 15.5

1905	1906	1907	1908	1909	1910	1911	1912	1913	1914
10·56	10·90	12·75	12·77	11·30	11·72	11·43	12·70	13·94	13·65

With 1910 as base year calculate the price relatives in each year.

2. *Table 15.6* gives price relatives for the years 1950–1958, the base year being 1947. Recalculate the price relatives with 1954 as the base year.

Table 15.6

1950	1951	1952	1953	1954	1955	1956	1957	1958
108	116	119	120	120	120	122	126	129

3. The Ministry of Labours' (1947) Interim Index of Retail Prices is given in *Table 15.7*.

Table 15.7

Group	Item	Weight
I	Food	348
II	Rent and Rates	88
III	Clothing	97
IV	Fuel and Light	65
V	Household goods	71
VI	Miscellaneous	35
VII	Services	79
VIII	Drink and tobacco	217
		1000

The weights quoted here are proportional to the total outlay on each item. Using the following price relatives calculate the appropriate index number. Food 114, Rent and Rates 128, Clothing 120, Fuel and Light 135, Household Goods 130, Miscellaneous 100, Services 140, Drink and Tobacco 150.

15.3. CRUDE AND STANDARDIZED DEATH RATES

The *crude death rate* for a given town or district is the number of deaths per 1000 population.

$$\text{Crude death rate} = \frac{\text{Total number of deaths}}{\text{Total population}} \times 1000 \quad \ldots (15.4)$$

As the measure takes no account of the numbers in different age groups it can be affected by a preponderance of any one group. It is not suitable as a comparison of the death rate between different places.

Example 15.7. Calculate the crude death rate for the two towns given in *Table 15.8*. Comment on the results.

Table 15.8

	Total population	Total deaths in 1950
XYZ Spa	25,000	800
ABC Industrial Town	120,000	1984

Crude death rate for XYZ spa $= \dfrac{800}{25,000} \times 1000 = 32 \cdot 0$

Crude death rate for ABC industrial town $= \dfrac{984}{120,000} \times 1000 = 8 \cdot 2$

Comment: it appears on a superficial glance that the industrial town is a much healthier place to live in. However, older people tend to retire to places like XYZ spa and thus the spa would have more in the older age groups than the industrial town. The older age groups having, naturally, a larger death rate, the crude death rate for the spa would tend to be larger.

To obtain a standardized death rate a *standard population* is used as a basis for comparison. As the mortality rates for males and females differ considerably the distribution of the population is given in two parts. Standardized death rates in Great Britain for the year 1946 and onwards have been calculated on the basis of the distribution by age and sex of the whole population at the year of observation.

To calculate the standardized death rate the crude death rates are found for each group under consideration. Using these crude death rates the total deaths per thousand of the standard population are found (crude death rate × standard population) and the sum of these is the standardized death rate.

Example 15.8. (See *Table 15.9*).

Table 15.9

Age range	Population	No. of deaths	Crude death rate	Standard population	Deaths per 1000 of standard population)
0–14	5000	60	12·0	325	3·90
15–34	8000	56	7·0	357	2·50
35–59	6000	48	8·0	243	1·94
60+	2000	135	67·5	75	5·06
				1000	13·40

Standardized death rate 13·40

Exercises 15c

1. Complete *Table 15.10* and state (*a*) the crude death rate, (*b*) the standardized death rate.

Table 15.10

Age range	Population	No. of deaths	Crude death rate	Standard population	Deaths (per 1000 of standard population)
Males					
0–4	3050	44		59	
5–14	4927	9		109	
15–34	9823	27		177	
35–59	8756	87		121	
60+	4805	299		34	
Females					
0–4	2998	42		55	
5–14	4876	10		102	
15–34	9836	26		180	
35–59	9112	84		122	
60+	6402	444		41	
				1000	

2. Using the figures given in *Table 15.11* calculate for any town (*a*) the crude rate of unemployment, (*b*) the standard rate.

Table 15.11

Age	Population distribution (Any town)	Population distribution standard population	No. unemployed
15–30	5000	300	100
30–45	6000	350	180
45–60	6000	250	480
60+	3000	100	660
Total	20,000	1000	

15.4. INTRODUCTION TO TIME SERIES

A *time series* is a set of observations taken at different times, usually at equal intervals. The observations can be plotted against time (see *Figure 15.1*).

Examples of time series are the total annual exports, the monthly unemployment rate, the daily average temperature, etc.

The variations in the set of observations are generally taken to be due to four factors:

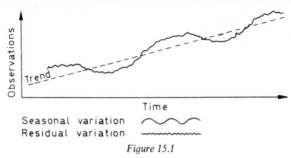

Figure 15.1

(*a*) long term trends,
(*b*) seasonal variations (short cyclical variations),
(*c*) long term cyclical movements,
(*d*) erratic variations.
For the purposes of this book the last two effects are combined and only long term trends (or trend), seasonal variations and residual variations are considered.

Long Term Trends (Secular Trend)

The trend of a time series is the way in which the graph appears to be going over a long interval of time. In *Figure 15.1* the trend curve is shown as a dotted line. Its general rise or fall is due to factors which only change slowly over a long period of time, e.g. population, technological improvements, etc.

Seasonal Variations

A regular rise and fall over specified intervals of time is referred to as a *seasonal variation*. The interval of time can be of any length, hours, days, weeks, years, etc. and the variations are of a cyclical type with a fairly definite period, e.g. rises in the number of goods sold just before Xmas, rises in the demand for Gas or Electricity at certain times during the day, and the number of customers using a restaurant during a day.

Residual Variations

Any other variation which cannot be ascribed to seasonal or long term variations is taken as happening entirely at random due to unpredictable causes, e.g. showers of rain, earthquakes, fires, etc. and is the *Residual Variation*.

Thus the overall variation of a time series is due to a function of these three factors: trend (T), seasonal variation (S) and residual variation (R). Either of the two simple functions defined below is used in the analysis of time series:

$$\text{Overall variation} = T + S + R \qquad \ldots\ldots(15.5)$$
$$\text{Overall variation} = T \times S \times R \qquad \ldots\ldots(15.6)$$

Examples 15.11 and 15.12 illustrate respectively the use of both of these expressions.

15.5. MOVING AVERAGES

Given a set of numbers $x_1, x_2, \ldots x_{n+1}, x_{n+2}, \ldots$ the *moving average of order n* is given by the following sequence of arithmetic means:

$$\frac{x_1 + x_2 + \ldots x_n}{n}, \quad \frac{x_2 + x_3 + \ldots x_{n+1}}{n}, \quad \frac{x_3 + x_4 + \ldots x_{n+2}}{n}$$

Example 15.9. Given the numbers 7, 7, 6, 6, 11, 11, 10, find the moving averages of order 4.

The moving averages are given by

$$\frac{7+7+6+6}{4}, \quad \frac{7+6+6+11}{4}, \quad \frac{6+6+11+11}{4}, \quad \frac{6+11+11+10}{4}$$

$$6{\cdot}5 \qquad\qquad 7{\cdot}5 \qquad\qquad 8{\cdot}5 \qquad\qquad 9{\cdot}5$$

If the data is given, for example, monthly, the average is known as the n month moving average. The moving average is centred at the middle of the period to which it refers. Note the position of the 6 month moving average in the next example.

Example 15.10. Find the 3 month and the 6 month moving average for the data given in *Table 15.12*.

Table 15.12

Month	Jan.	Feb.	March	April	May	June	July	Aug.	Sept.	Oct.	Nov.	Dec.
No. of workmen	51	57	63	68	70	72	75	81	87	92	95	96

In *Table 15.13* the moving total column has been included for ease of calculation.

The importance of moving averages is that they enable us to average out a seasonal variation if its period is known or can be deduced. Examples 15.9 and 15.10 were constructed to illustrate this point. In

Table 15.13

Month	No. of workmen	3 month moving total	3 month moving average	6 month moving total	6 month moving average
Jan.	51				
Feb.	58	172	57·3		
Mar.	63	189	63·0	382	63·7
Apr.	68	201	67·0	406	67·7
May	70	210	70·0	429	71·5
June	72	217	72·3	453	75·5
July	75	228	76·0	477	79·5
Aug.	81	243	81·0	502	83·7
Sept.	87	260	86·7	526	87·7
Oct.	92	274	91·3		
Nov.	95	283	94·3		
Dec.	96				

Example 15.9, where a simple trend 5, 6, 7, 8, 9, 10, 11, 12 was super-imposed on a cyclical variation 2, 1, -1, -2, 2, 1, -1, -2, a moving average of order 4 (the same order as the cyclical variation) isolated the trend.

It is important that the *correct order* of moving average is used. In *Figure 15.2* both the 3 month and 6 month moving average, obtained in Example 15.10, are plotted. The 6 month moving average eliminates more of the cyclical variation than does the 3 month moving average.

15.6. ANALYSIS OF A TIME SERIES

Our aim is to separate the variation into the three components trend, seasonal variation, and residual. First the order n of the seasonal variation is obtained from knowledge of the circumstances in which the data arose. A moving average of the nth order is then obtained. The trend is then subtracted from the original figures giving seasonal variations *plus* residual. These composite values will cycle in blocks of n (see Example 15.11 cols. vi–viii) giving several values for each part of the cycle. We then average the corresponding values in each cycle assuming that in this way random variations will tend to cancel out. Before using these average values as estimates of the seasonal variations they are adjusted so that their sum is zero (see footnote to Example 15.11).

Example 15.11. Table 15.14 gives the analysis of a time series whose period of seasonal variation was one year.

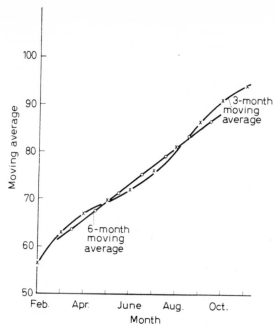

Figure 15.2. To illustrate the data of Example 15.10

Notes on the calculations shown in Example 15.11

(*a*) Columns (iii) and (iv) are centred between the original values because we are averaging four values, *hence*:

(*b*) Column (v) is the average of two consecutive values of the moving average so that it can appear opposite the original values.

(*c*) Columns (vi)–(ix) give the differences between column (ii) and column (v). That is the difference between the original values and the trend and hence represents the seasonal variation + residual.

(*d*) Columns (vi)–(ix) are totalled and averaged; this is assumed to eliminate random variations and give a good estimate of the quarterly variations.

(*e*) The sum of the quarterly variations should be zero and the adjustment is made in the footnote.

We obtain $-1\cdot78$, $+2\cdot20$, $-1\cdot97$, $+1\cdot57$ as the average quarterly deviations. That is the original values for the first quarter have been decreased by $1\cdot78$ due to seasonal change, the values for the second quarter increased by $2\cdot20$ etc. We apply these average quarterly deviations as the *corrections* to the original production figures which give column (x).

Table 15.14

| Quarter (i) | Total cost (ii) | 4-quarter moving total (iii) | 4-quarter moving average (iv) | Mean value (trend) (v) | Deviations (seasonal variation + residual) | | | | Trend + residual Col. ii – corrected estimates of quarterly variations (x) | Residual variation (col. x – col. v) (xi) |
					1st quarter (vi)	2nd quarter (vii)	3rd quarter (viii)	4th quarter (ix)		
1950 I	33·4									
II	38·3	145·0	36·250							
III	34·3	146·9	36·725	36·49			−2·19			
IV	39·0	148·2	37·050	36·89				+2·11		
1951 I	35·3	149·1	37·275	37·16	−1·86				37·08	−0·08
II	39·6	149·2	37·30	37·29		+2·31			37·40	+0·11
III	35·2	150·9	37·725	37·51			−2·31		37·17	−0·34
III	39·1	152·5	38·125	37·93				+1·17	37·53	−0·40
IV	37·0	155·8	38·95	38·54	−1·54				38·78	+0·24
1952 I	41·2	159·5	39·875	39·41		+1·79			39·00	−0·41
II	38·5	161·7	40·425	40·15			−1·65		40·47	+0·32
III	42·8	163·3	40·825	40·63				+2·17	41·23	+0·60
IV										

Year	Quarter	Value	4-qtr moving total	Moving average	Centered moving average	I	II	III	IV		Deviation
1953	I	39·2			40·88	−1·68				40·98	+0·10
	II	42·8	163·7	40·925	40·86		+1·94			40·60	−0·20
	III	38·9	163·2	40·80	40·78			−1·88		40·87	+0·09
	IV	42·3	163·0	40·75	40·78				+1·52	40·73	−0·05
1954	I	39·0	163·2	40·80	40·89	−1·89				40·78	−0·11
	II	43·0	163·9	40·975	41·06		+1·94			40·80	+0·74
	III	39·6	164·6	41·15	41·39			−1·79		41·57	+0·18
	IV	43·0	166·5	41·625	42·09				+0·91	41·43	−0·66
1955	I	40·9	170·2	42·55	42·78	−1·88					
	II	46·7	172·0	43·00	43·64		+3·06				
	III	41·4	177·1	44·275							
	IV	48·1									
					Totals	−8·85	+11·04	−9·82	+7·88		
				Averages (estimate of quarterly variation)		−1·77	+2·21	−1·96	+1·58		

The grand total of the Deviations should be zero. The grand total of the above is +0·06. Therefore we apply a correction of −0·06 ÷ 4 = −0·01.
Corrected estimates of quarterly variations −1·78 +2·20 −1·97 +1·57.

The above analysis has been based on the supposition that the overall variation is equal to the sum of the variations due to trend (T) seasonal (S) and residual (R), that is, we used equation (15.5). As stated in section 15·7, an alternative is to regard the overall variation as the product of these three components, that is

$$\text{Overall variation} = T \times S \times R \text{ as in equation (15.6)}$$

In this case the seasonal and residual variations are regarded as a *proportion* of the grand trend and not as fixed amounts. To illustrate this idea consider a store whose gross sales are £10,000 per week in January with a maximum seasonal variation of $\pm$£1000 (10 per cent). If, in the course of the year the store prospered and the gross sales rose to £20,000 per week, then if the second model ($T \times S \times R$) is accepted we should expect the seasonal variation to remain at 10 per cent, i.e. become $\pm$£2000 in the next year rather than remain at $\pm$£1000. This is a fairly extreme case as not many variables double in the course of a year.

The analysis of the time series when the overall variation is taken as the product of the three components is exactly the same as in Example 15.11, only we work with logs of the data. The reason for this is that if

$$\text{Overall variation } = T \times S \times R$$

then

$$\text{Log (overall variation)} = \log T + \log S + \log R$$

Example 15.12. Table 15.15 gives, in £1000's, the quarterly sales of a store; assuming an overall variation of $T \times S \times R$, analyse the data.

Table 15.15

	1950	1951	1952	1953	1954	1955
1st quarter	33·4	35·3	37·0	39·2	39·0	40·9
2nd quarter	38·3	39·6	41·2	42·8	43·0	46·7
3rd quarter	34·3	35·2	38·5	38·9	39·6	41·4
4th quarter	39·0	39·1	42·8	42·3	43·0	48·1

We first find the logarithms of all the given sales figures, making a similar table (*Table 15.16*).

These figures are now analysed, as in Example 15.11, in the same sort of table and the final results are antilogged.

Table 15.16

	1950	1951	1952	1953	1954	1955
1st quarter	1·5237	1·5478	1·5682	1·5933	1·5911	1·6117*
2nd quarter	1·5832	1·5977	1·6149	1·6314	1·6335	1·6693
3rd quarter	1·5353	1·5465	1·5855	1·5899	1·5977	1·6170
4th quarter	1·5911	1·5922	1·6314	1·6263	1·6335	1·6821

Exercises 15d

1. Draw a frequency polygon to illustrate the data of Example 15.10. On the same diagram draw the graph of the 6 month moving average.

2. *Table 15.17* gives (in thousands) the number of registered unemployed for each month for the years 1940–1942. Calculate the

Table 15.17

	Jan.	Feb.	March	April	May	June	July	Aug.	Sept.	Oct.	Nov.	Dec.
1940	1471	1466	1083	931	843	709	755	721	789	797	768	683
1941	653	528	391	343	317	252	234	214	178	171	159	151
1942	162	161	137	119	114	106	103	115	109	100	104	100

12 month moving average and draw a graph of the average. On the same diagram (with the same axes) draw a graph of the data (work to the nearest whole number).

3. Calculate the 3 day and the 5 day moving average for the data given in *Table 15.18*.

Table 15.18

Mon.	Tues.	Wed.	Th.	Fri.	Sat.	Sun.	Mon.	Tues.	Wed.	Th.
31	23	19	21	21	15	4	8	9	6	0

4. Re-organize the data in question 2 to give the number of registered unemployed (in thousands) for each of the four quarters of the years 1940, 1941, 1942. Estimate the seasonal variation in unemployment for each of the four quarters of the year (use a four-quarter moving average and work to the nearest whole number).

5. In Example 15.11 the four-quarter moving average was centred

between the successive quarterly figures [see column (iii)]. In column (iv) we calculated the average of each two successive values of this moving average in order to obtain values opposite the successive quarterly figures. Show that this procedure is the same as taking a 5 year weighted moving average with weights 1, 2, 2, 2, 1.

6. The number of items (in thousands) of a certain product which were sold in a given town are given in *Table 15.19*.

Table 15.19

Year	Amount	Year	Amount
1940	392	1950	443
1941	596	1951	644
1942	553	1952	607
1943	227	1953	276
1944	210	1954	260
1945	415	1955	465
1946	620	1956	673
1947	574	1957	627
1948	252	1958	301
1949	238	1959	298

Plot a scatter diagram and decide the length of the seasonal variation. Hence analyse the data fully into the variation due to trend, seasonal variation, and residual variation.

EXERCISES 15

1. Two chemists each test a bulk delivery of chemicals for impurities.

The first one makes 12 analyses and finds an average impurity of 2·41 per cent. The second makes 15 analyses and finds an average impurity of 2·80 per cent. If the two chemists are equally reliable in their work what is the best estimate of the average impurity from the two sets of results?

2. The retail price of petrol varies according to the area in which it is sold. *Table 15.20* gives the relative amounts of petrol sold at the different prices. Find the average price of all the petrol sold.

Table 15.20

Pence per gallon	Proportion
62·0	0·35
62·5	0·52
63·0	0·13
	1·00

3. *Table 15.21* gives the quarterly figures of the average percentage of insured workers unemployed in the building industries of Great Britain.

Table 15.21

Year	1927	1928	1929	1930	1931
1st quarter	12·1	14·5	19·1	17·5	24·9
2nd quarter	6·8	9·7	9·0	12·9	18·2
3rd quarter	7·6	10·6	9·3	14·5	19·1
4th quarter	13·4	14·2	15·2	20·4	26·5

Calculate a four-quarter moving average and draw graphs using the same axes of the data and the four-quarter moving average (work correct to one place of decimals).

4. The figures in *Table 15.22* refer to the average percentage of insured workers unemployed in the building industries of Great Britain.

Table 15.22

(a)				(b)	
Year	1932	1933		Period	Four quarter moving average
1st quarter	31·3			1st quarter 1932–4th quarter 1932	28·950
2nd quarter	26·3			2nd quarter 1932–1st quarter 1933	29·325
3rd quarter	27·6			3rd quarter 1932–2nd quarter 1933	27·950
4th quarter				4th quarter 1932–3rd quarter 1933	25·875
				1st quarter 1933–4th quarter 1933	23·725

Complete *Table (a)* using *Table (b)*.

5. The figures in *Table 15.23* give the variation in the wholesale price of eggs (in shillings per 120 to the nearest 0·1 of a shilling). Using a 12 month moving average, fit a trend to this data. On the same axes draw a graph of the trend and of the original data.

Table 15.23

Month	1936	1937	1938
Jan.	15·4	12·9	16·2
Feb.	14·2	13·7	14·5
March	9·6	10·8	10·3
April	8·5	9·0	10·3
May	9·5	9·7	11·3
June	10·5	11·8	12·2
July	12·7	14·5	15·3
Aug.	15·0	15·9	16·8
Sept.	15·5	17·7	18·4
Oct.	20·4	19·8	19·9
Nov.	20·5	23·3	21·8
Dec.	18·8	21·3	17·7

6. Calculate from the figures in *Table 15.24* the standardized death rate for all the females in Town A using the total population of England and Wales as a standard population (work correct to two places of decimals).

Table 15.24

Age group	Age distribution of females in England and Wales	Female population (Town A)	No. deaths of females (Town A)
0–4	64	1215	16
5–14	138	1816	10
15–24	155	1931	10
25–34	165	2035	28
35–44	147	800	9
45–64	234	950	18
65–	97	120	4
	1000	8867	

7. Using the data and the results of question 3, find the quarterly variations for unemployed building workers in Great Britain (see Example 15.11) (work correct to one place of decimals).

8. (a) Calculate the geometric mean of the following figures: 0·53, 2·79, 3·84, 12·75, (b) using weights of 5, 2, 3, 1 respectively, calculate the weighted geometric mean.

9. From *Table 15.25* using 1950 as a base year and weights as indicated, calculate the index numbers for the years 1951 and 1952 (work to one place of decimals).

Table 15.25

| Commodity | Weight | 1950 | Price per unit | |
			1951	1952
A	21	26·0	29·0	25·0
B	3	33·1	36·3	34·7
C	3	115·7	120·7	116·2
D	2	21·3	19·2	22·4
E	12	52·6	52·7	48·8

10. Using the data of question 9 construct a table showing the price relatives for the five different commodities in the different years. Take 1950 prices as a base. From your table of price relatives calculate index numbers for 1951 and 1952 using as weight the total expenditures in 1950 on the five different commodities. Check your answers with those of question 9 (work correct to four significant figures).

11. (a) A sequence has 24 numbers. How many moving averages of order five will there be? Repeat the question for (b) 50, (c) 130 numbers.

Obtain the general formula for a sequence of S numbers using a moving average of order A.

12. The average price of wheat and the quantities sold at four markets are given in *Table 15.26*.

Table 15.26

Market	Average price (per quarter)		Quantity sold (quarters)
	s.	d.	
A	27	3	36,000
B	28	8	1000
C	29	1	16,000
D	27	2	12,000

Find the mean price per quarter for the four markets, weighting each price with the quantity sold.

Would it be possible for the average price at each of the four markets to rise and yet the weighted mean price to fall. If so, under what conditions would this happen?

13. *Table 15.27* gives the quarterly figures of the average percentage of unemployed building workers:

Table 15.27

Year	Quarter			
	I	II	III	IV
1934	22·9	15·6	16·1	19·3
1935	21·2	14·2	14·1	16·5
1936	20·8	11·2	11·3	14·9

Obtain the average seasonal variations and the residual variations (see Example 15.11 and work to two places of decimals).

14. It is thought that the data given in *Table 15.28* have a general trend and a 5 year seasonal variation. Using a 5 year moving average analyse the data and see if you think that this is in fact so.

Table 15.28

1950	1951	1952	1953	1954	1955	1956	1957	1958	1959	1960	1961	1962	1963
29	44	50	50	45	74	77	85	104	117	114	129	155	165

APPENDIX

Table 1. Poisson Probabilities. $e^{-a}a^x/x!$ for values of x 0(1)8, and for values 0·1(0·1)5·0 of the mean a

mean a	0	1	2	3	4	5	6	7	8
0·1	0·9048	0·0905	0·0045	0·0002	0·0000	0·0000	0·0000	0·0000	0·0000
0·2	0·8187	0·1637	0·0164	0·0011	0·0001	0·0000	0·0000	0·0000	0·0000
0·3	0·7408	0·2222	0·0333	0·0033	0·0003	0·0000	0·0000	0·0000	0·0000
0·4	0·6703	0·2681	0·0536	0·0072	0·0007	0·0001	0·0000	0·0000	0·0000
0·5	0·6065	0·3033	0·0758	0·0126	0·0016	0·0002	0·0000	0·0000	0·0000
0·6	0·5488	0·3293	0·0988	0·0198	0·0030	0·0004	0·0000	0·0000	0·0000
0·7	0·4966	0·3476	0·1217	0·0284	0·0050	0·0007	0·0001	0·0000	0·0000
0·8	0·4493	0·3595	0·1438	0·0383	0·0077	0·0012	0·0002	0·0000	0·0000
0·9	0·4066	0·3659	0·1647	0·0494	0·0111	0·0020	0·0003	0·0000	0·0000
1·0	0·3679	0·3679	0·1839	0·0613	0·0153	0·0031	0·0005	0·0001	0·0000
1·1	0·3329	0·3662	0·2014	0·0738	0·0203	0·0045	0·0008	0·0001	0·0000
1·2	0·3012	0·3614	0·2169	0·0867	0·0260	0·0062	0·0012	0·0002	0·0000
1·3	0·2725	0·3543	0·2303	0·0998	0·0324	0·0084	0·0018	0·0003	0·0001
1·4	0·2466	0·3452	0·2417	0·1128	0·0395	0·0111	0·0026	0·0005	0·0001
1·5	0·2231	0·3347	0·2510	0·1255	0·0471	0·0141	0·0035	0·0008	0·0001
1·6	0·2019	0·3230	0·2584	0·1378	0·0551	0·0176	0·0047	0·0011	0·0002
1·7	0·1827	0·3106	0·2640	0·1496	0·0636	0·0216	0·0061	0·0015	0·0003
1·8	0·1653	0·2975	0·2678	0·1607	0·0723	0·0260	0·0078	0·0020	0·0005
1·9	0·1496	0·2842	0·2700	0·1710	0·0812	0·0309	0·0098	0·0027	0·0006
2·0	0·1353	0·2707	0·2707	0·1804	0·0902	0·0361	0·0120	0·0034	0·0009
2·1	0·1225	0·2572	0·2700	0·1890	0·0992	0·0417	0·0146	0·0044	0·0011
2·2	0·1108	0·2438	0·2681	0·1966	0·1082	0·0476	0·0174	0·0055	0·0015
2·3	0·1003	0·2306	0·2652	0·2033	0·1169	0·0538	0·0206	0·0068	0·0019
2·4	0·0907	0·2177	0·2613	0·2090	0·1254	0·0602	0·0241	0·0083	0·0025
2·5	0·0821	0·2052	0·2565	0·2138	0·1336	0·0668	0·0278	0·0099	0·0031
2·6	0·0743	0·1931	0·2510	0·2176	0·1414	0·0735	0·0319	0·0118	0·0038
2·7	0·0672	0·1815	0·2450	0·2205	0·1488	0·0804	0·0362	0·0139	0·0047
2·8	0·0608	0·1703	0·2384	0·2225	0·1557	0·0872	0·0407	0·0163	0·0057
2·9	0·0550	0·1596	0·2314	0·2237	0·1622	0·0940	0·0455	0·0188	0·0068
3·0	0·0498	0·1494	0·2240	0·2240	0·1680	0·1008	0·0504	0·0216	0·0081
3·1	0·0450	0·1397	0·2165	0·2237	0·1733	0·1075	0·0555	0·0246	0·0095
3·2	0·0408	0·1304	0·2087	0·2226	0·1781	0·1140	0·0608	0·0278	0·0111
3·3	0·0369	0·1217	0·2008	0·2209	0·1823	0·1203	0·0662	0·0312	0·0129
3·4	0·0334	0·1135	0·1929	0·2186	0·1858	0·1264	0·0716	0·0348	0·0148
3·5	0·0302	0·1057	0·1850	0·2158	0·1888	0·1322	0·0771	0·0385	0·0169
3·6	0·0273	0·0984	0·1771	0·2125	0·1912	0·1377	0·0826	0·0425	0·0191
3·7	0·0247	0·0915	0·1692	0·2087	0·1931	0·1429	0·0881	0·0466	0·0215
3·8	0·0224	0·0850	0·1615	0·2046	0·1944	0·1477	0·0936	0·0508	0·0241
3·9	0·0202	0·0789	0·1539	0·2001	0·1951	0·1522	0·0989	0·0551	0·0269
4·0	0·0183	0·0733	0·1465	0·1954	0·1954	0·1563	0·1042	0·0595	0·0298
4·1	0·0166	0·0679	0·1393	0·1904	0·1951	0·1600	0·1093	0·0640	0·0328

Table 1—continued

mean					x				
a	0	1	2	3	4	5	6	7	8
4·2	0·0150	0·0630	0·1323	0·1852	0·1944	0·1633	0·1143	0·0686	0·0360
4·3	0·0136	0·0583	0·1254	0·1798	0·1933	0·1662	0·1191	0·0732	0·0393
4·4	0·0123	0·0540	0·1188	0·1743	0·1917	0·1687	0·1237	0·0778	0·0428
4·5	0·0111	0·0500	0·1125	0·1687	0·1898	0·1708	0·1281	0·0824	0·0463
4·6	0·0101	0·0462	0·1063	0·1631	0·1875	0·1725	0·1323	0·0869	0·0500
4·7	0·0091	0·0427	0·1005	0·1574	0·1849	0·1738	0·1362	0·0914	0·0537
4·8	0·0082	0·0395	0·0948	0·1517	0·1820	0·1747	0·1398	0·0959	0·0575
4·9	0·0074	0·0365	0·0894	0·1460	0·1789	0·1753	0·1432	0·1002	0·0614
5·0	0·0067	0·0337	0·0842	0·1404	0·1755	0·1755	0·1462	0·1044	0·0653

Table 2. Binomial Probabilities. The table gives (in columns) the value of $^nC_x(1 - p)^{n-x}p^x$
[the $(x + 1)$th term in the expansion of $(q + p)^n$], for values of p 0·1(0·1)0·9, of n 3(1)10 and of x 0(1) n.

					$N = 3$				
x	P = 0·1	P = 0·2	P = 0·3	P = 0·4	P = 0·5	P = 0·6	P = 0·7	P = 0·8	P = 0·9
0	0·7290	0·5120	0·3430	0·2160	0·1250	0·0640	0·0270	0·0080	0·0010
1	0·2430	0·3840	0·4410	0·4320	0·3750	0·2880	0·1890	0·0960	0·0270
2	0·0270	0·0960	0·1890	0·2880	0·3750	0·4320	0·4410	0·3840	0·2430
3	0·0010	0·0080	0·0270	0·0640	0·1250	0·2160	0·3430	0·5120	0·7290

					$N = 4$				
x	P = 0·1	P = 0·2	P = 0·3	P = 0·4	P = 0·5	P = 0·6	P = 0·7	P = 0·8	P = 0·9
0	0·6561	0·4096	0·2401	0·1296	0·0625	0·0256	0·0081	0·0016	0·0001
1	0·2916	0·4096	0·4116	0·3456	0·2500	0·1536	0·0756	0·0256	0·0036
2	0·0486	0·1536	0·2646	0·3456	0·3750	0·3456	0·2646	0·1536	0·0486
3	0·0036	0·0256	0·0756	0·1536	0·2500	0·3456	0·4116	0·4096	0·2916
4	0·0001	0·0016	0·0081	0·0256	0·0625	0·1296	0·2401	0·4096	0·6561

					$N = 5$				
x	P = 0·1	P = 0·2	P = 0·3	P = 0·4	P = 0·5	P = 0·6	P = 0·7	P = 0·8	P = 0·9
0	0·5905	0·3277	0·1681	0·0778	0·0313	0·0102	0·0024	0·0003	0·0000
1	0·3281	0·4096	0·3602	0·2592	0·1563	0·0768	0·0284	0·0064	0·0004
2	0·0729	0·2048	0·3087	0·3456	0·3125	0·2304	0·1323	0·0512	0·0081
3	0·0081	0·0512	0·1323	0·2304	0·3125	0·3456	0·3087	0·2048	0·0729
4	0·0004	0·0064	0·0283	0·0768	0·1563	0·2592	0·3602	0·4096	0·3280
5	0·0000	0·0003	0·0024	0·0102	0·0313	0·0778	0·1681	0·3277	0·5905

					$N = 6$				
x	P = 0·1	P = 0·2	P = 0·3	P = 0·4	P = 0·5	P = 0·6	P = 0·7	P = 0·8	P = 0·9
0	0·5314	0·2621	0·1176	0·0467	0·0156	0·0041	0·0007	0·0001	0·0000
1	0·3543	0·3932	0·3025	0·1866	0·0938	0·0369	0·0102	0·0015	0·0001

Table 2—continued

				$N = 6$					
x	$P = 0\cdot1$	$P = 0\cdot2$	$P = 0\cdot3$	$P = 0\cdot4$	$P = 0\cdot5$	$P = 0\cdot6$	$P = 0\cdot7$	$P = 0\cdot8$	$P = 0\cdot9$
2	0·0984	0·2458	0·3241	0·3110	0·2344	0·1382	0·0595	0·0154	0·0012
3	0·0146	0·0819	0·1852	0·2765	0·3125	0·2765	0·1852	0·0819	0·0146
4	0·0012	0·0154	0·0595	0·1382	0·2344	0·3110	0·3241	0·2458	0·0984
5	0·0001	0·0015	0·0102	0·0369	0·0938	0·1866	0·3025	0·3932	0·3543
6	0·0000	0·0001	0·0007	0·0041	0·0156	0·0467	0·1176	0·2621	0·5314

				$N = 7$					
x	$P = 0\cdot1$	$P = 0\cdot2$	$P = 0\cdot3$	$P = 0\cdot4$	$P = 0\cdot5$	$P = 0\cdot6$	$P = 0\cdot7$	$P = 0\cdot8$	$P = 0\cdot9$
0	0·4783	0·2097	0·0824	0·0280	0·0078	0·0016	0·0002	0·0000	0·0000
1	0·3720	0·3670	0·2471	0·1306	0·0547	0·0172	0·0036	0·0004	0·0000
2	0·1240	0·2753	0·3177	0·2613	0·1641	0·0774	0·0250	0·0043	0·0002
3	0·0230	0·1147	0·2269	0·2903	0·2734	0·1935	0·0972	0·0287	0·0026
4	0·0026	0·0287	0·0972	0·1935	0·2734	0·2903	0·2269	0·1147	0·0230
5	0·0002	0·0043	0·0250	0·0774	0·1641	0·2613	0·3177	0·2753	0·1240
6	0·0000	0·0004	0·0036	0·0172	0·0547	0·1306	0·2471	0·3670	0·3720
7	0·0000	0·0000	0·0002	0·0016	0·0078	0·0280	0·0824	0·2097	0·4783

				$N = 8$					
x	$P = 0\cdot1$	$P = 0\cdot2$	$P = 0\cdot3$	$P = 0\cdot4$	$P = 0\cdot5$	$P = 0\cdot6$	$P = 0\cdot7$	$P = 0\cdot8$	$P = 0\cdot9$
0	0·4305	0·1678	0·0576	0·0168	0·0039	0·0007	0·0001	0·0000	0·0000
1	0·3826	0·3355	0·1977	0·0896	0·0313	0·0079	0·0012	0·0001	0·0000
2	0·1488	0·2936	0·2965	0·2090	0·1094	0·0413	0·0100	0·0011	0·0000
3	0·0331	0·1468	0·2541	0·2787	0·2188	0·1239	0·0467	0·0092	0·0004
4	0·0046	0·0459	0·1361	0·2322	0·2734	0·2322	0·1361	0·0459	0·0046
5	0·0004	0·0092	0·0467	0·1239	0·2188	0·2787	0·2541	0·1468	0·0331
6	0·0000	0·0011	0·0100	0·0413	0·1094	0·2090	0·2965	0·2936	0·1488
7	0·0000	0·0001	0·0012	0·0079	0·0313	0·0896	0·1977	0·3355	0·3826
8	0·0000	0·0000	0·0001	0·0007	0·0039	0·0168	0·0576	0·1678	0·4305

				$N = 9$					
x	$P = 0\cdot1$	$P = 0\cdot2$	$P = 0\cdot3$	$P = 0\cdot4$	$P = 0\cdot5$	$P = 0\cdot6$	$P = 0\cdot7$	$P = 0\cdot8$	$P = 0\cdot9$
0	0·3874	0·1342	0·0404	0·0101	0·0020	0·0003	0·0000	0·0000	0·0000
1	0·3874	0·3020	0·1556	0·0605	0·0176	0·0035	0·0004	0·0000	0·0000
2	0·1722	0·3020	0·2668	0·1612	0·0703	0·0212	0·0039	0·0003	0·0000
3	0·0446	0·1762	0·2668	0·2508	0·1641	0·0743	0·0210	0·0028	0·0001
4	0·0074	0·0661	0·1715	0·2508	0·2461	0·1672	0·0735	0·0165	0·0008
5	0·0008	0·0165	0·0735	0·1672	0·2461	0·2508	0·1715	0·0661	0·0074
6	0·0001	0·0028	0·0210	0·0743	0·1641	0·2508	0·2668	0·1762	0·0446
7	0·0000	0·0003	0·0039	0·0212	0·0703	0·1612	0·2668	0·3020	0·1722
8	0·0000	0·0000	0·0004	0·0035	0·0176	0·0605	0·1556	0·3020	0·3874
9	0·0000	0·0000	0·0000	0·0003	0·0020	0·0101	0·0404	0·1342	0·3874

				$N = 10$					
x	$P = 0\cdot1$	$P = 0\cdot2$	$P = 0\cdot3$	$P = 0\cdot4$	$P = 0\cdot5$	$P = 0\cdot6$	$P = 0\cdot7$	$P = 0\cdot8$	$P = 0\cdot9$
0	0·3487	0·1074	0·0282	0·0060	0·0010	0·0001	0·0000	0·0000	0·0000

Table 2—continued

x	N = 10								
	P = 0·1	P = 0·2	P = 0·3	P = 0·4	P = 0·5	P = 0·6	P = 0·7	P = 0·8	P = 0·9
1	0·3874	0·2684	0·1211	0·0403	0·0098	0·0016	0·0001	0·0000	0·0000
2	0·1937	0·3020	0·2335	0·1209	0·0439	0·0106	0·0014	0·0001	0·0000
3	0·0574	0·2013	0·2668	0·2150	0·1172	0·0425	0·0090	0·0008	0·0000
4	0·0112	0·0881	0·2001	0·2508	0·2051	0·1115	0·0368	0·0055	0·0001
5	0·0015	0·0264	0·1029	0·2007	0·2461	0·2007	0·1029	0·0264	0·0015
6	0·0001	0·0055	0·0368	0·1115	0·2051	0·2508	0·2001	0·0881	0·0112
7	0·0000	0·0008	0·0090	0·0425	0·1172	0·2150	0·2668	0·2013	0·0574
8	0·0000	0·0001	0·0014	0·0106	0·0439	0·1209	0·2335	0·3020	0·1937
9	0·0000	0·0000	0·0001	0·0016	0·0098	0·0403	0·1211	0·2684	0·3874
10	0·0000	0·0000	0·0000	0·0001	0·0010	0·0060	0·0282	0·1074	0·3487

Table 3. Random Numbers. Each digit is an independent sample from a population in which the digits 0, 1, 2, ... 9 are equally likely, thus each one has a probability of $\frac{1}{10}$.

40	85	03	89	17	14	32	13	17	51	09	03	78	31	93	88
12	34	79	10	50	40	63	79	71	71	53	66	93	86	79	42
83	60	32	02	83	13	98	80	03	58	63	08	79	59	40	56
19	90	74	55	94	08	93	31	17	97	35	65	30	32	26	24
91	62	30	14	49	98	20	31	75	16	90	85	96	75	85	85
93	33	29	90	79	51	39	09	58	66	98	15	95	99	12	64
31	64	29	20	43	74	78	89	63	23	47	26	90	59	03	73
70	59	12	93	90	24	52	34	55	17	73	69	13	44	00	76
25	74	06	33	09	21	19	20	60	29	25	45	52	65	39	44
56	45	62	37	40	12	33	00	96	62	35	51	53	64	53	99
24	64	10	16	46	91	79	35	11	84	67	16	90	71	57	74
69	27	94	09	09	05	63	81	57	42	39	26	58	51	23	52
76	78	51	58	95	44	81	62	56	00	92	46	09	98	86	29
21	30	12	11	52	79	22	82	72	21	15	17	21	31	99	09
17	61	74	52	91	53	96	53	18	22	46	35	13	21	65	65
11	36	46	21	15	16	12	74	20	17	45	46	61	30	41	02

Table 4. The Normal Distribution Function

x	$\Phi(x)$	x	$\Phi(x)$	x	$\Phi(x)$	x	$\overset{*}{\Phi}(x)$	x	$\Phi(x)$
0·00	0·5000 $_{40}$	0·50	0·6915 $_{35}$	1·00	0·8413 $_{25}$	1·50	0·9332 $_{13}$	2·00	0·97725 $_{53}$
0·01	0·5040 $_{40}$	0·51	0·6950 $_{35}$	1·01	0·8438 $_{23}$	1·51	0·9345 $_{12}$	2·01	0·97778 $_{53}$
0·02	0·5080 $_{40}$	0·52	0·6985 $_{34}$	1·02	0·8461 $_{24}$	1·52	0·9357 $_{13}$	2·02	0·97831 $_{51}$
0·03	0·5120 $_{40}$	0·53	0·7019 $_{35}$	1·03	0·8485 $_{23}$	1·53	0·9370 $_{12}$	2·03	0·97882 $_{50}$
0·04	0·5160 $_{39}$	0·54	0·7054 $_{34}$	1·04	0·8508 $_{23}$	1·54	0·9382 $_{12}$	2·04	0·97932 $_{50}$
0·05	0·5199 $_{40}$	0·55	0·7088 $_{35}$	1·05	0·8531 $_{23}$	1·55	0·9394 $_{12}$	2·05	0·97982 $_{48}$
0·06	0·5329 $_{40}$	0·56	0·7123 $_{34}$	1·06	0·8554 $_{23}$	1·56	0·9406 $_{12}$	2·06	0·98030 $_{47}$
0·07	0·5279 $_{40}$	0·57	0·7157 $_{33}$	1·07	0·8577 $_{22}$	1·57	0·9418 $_{11}$	2·07	0·98077 $_{47}$
0·08	0·5319 $_{40}$	0·58	0·7190 $_{34}$	1·08	0·8599 $_{22}$	1·58	0·9429 $_{12}$	2·08	0·98124 $_{45}$
0·09	0·5359 $_{39}$	0·59	0·7224 $_{33}$	1·09	0·8621 $_{22}$	1·59	0·9441 $_{11}$	2·09	0·98169 $_{45}$
0·10	0·5398 $_{40}$	0·60	0·7257 $_{34}$	1·10	0·8643 $_{22}$	1·60	0·9452 $_{11}$	2·10	0·98214 $_{43}$
0·11	0·5438 $_{40}$	0·61	0·7291 $_{33}$	1·11	0·8665 $_{21}$	1·61	0·9463 $_{11}$	2·11	0·98257 $_{43}$
0·12	0·5478 $_{39}$	0·62	0·7324 $_{33}$	1·12	0·8686 $_{22}$	1·62	0·9474 $_{10}$	2·12	0·98300 $_{41}$
0·13	0·5517 $_{40}$	0·63	0·7357 $_{32}$	1·13	0·8708 $_{21}$	1·63	0·9484 $_{11}$	2·13	0·98341 $_{41}$
0·14	0·5557 $_{39}$	0·64	0·7389 $_{33}$	1·14	0·8729 $_{20}$	1·64	0·9495 $_{10}$	2·14	0·98382 $_{40}$
0·15	0·5596 $_{40}$	0·65	0·7422 $_{32}$	1·15	0·8749 $_{21}$	1·65	0·9505 $_{10}$	2·15	0·98422 $_{39}$
0·16	0·5636 $_{39}$	0·66	0·7454 $_{32}$	1·16	0·8770 $_{20}$	1·66	0·9515 $_{10}$	2·16	0·98461 $_{39}$
0·17	0·5675 $_{39}$	0·67	0·7486 $_{31}$	1·17	0·8790 $_{20}$	1·67	0·9525 $_{10}$	2·17	0·98500 $_{37}$
0·18	0·5714 $_{39}$	0·68	0·7517 $_{32}$	1·18	0·8810 $_{20}$	1·68	0·9535 $_{10}$	2·18	0·98537 $_{37}$
0·19	0·5753 $_{40}$	0·69	0·7549 $_{31}$	1·19	0·8830 $_{19}$	1·69	0·9545 $_{9}$	2·19	0·98574 $_{36}$
0·20	0·5793 $_{39}$	0·70	0·7580 $_{31}$	1·20	0·8849 $_{20}$	1·70	0·9554 $_{10}$	2·20	0·98610 $_{35}$
0·21	0·5832 $_{39}$	0·71	0·7611 $_{31}$	1·21	0·8869 $_{19}$	1·71	0·9564 $_{9}$	2·21	0·98645 $_{34}$
0·22	0·5871 $_{39}$	0·72	0·7642 $_{31}$	1·22	0·8888 $_{19}$	1·72	0·9573 $_{9}$	2·22	0·98679 $_{34}$
0·23	0·5910 $_{38}$	0·73	0·7673 $_{31}$	1·23	0·8907 $_{18}$	1·73	0·9582 $_{9}$	2·23	0·98713 $_{32}$
0·24	0·5948 $_{39}$	0·74	0·7704 $_{30}$	1·24	0·8925 $_{19}$	1·74	0·9591 $_{8}$	2·24	0·98745 $_{33}$
0·25	0·5987 $_{39}$	0·75	0·7734 $_{30}$	1·25	0·8944 $_{18}$	1·75	0·9599 $_{9}$	2·25	0·98778 $_{31}$
0·26	0·6026 $_{38}$	0·76	0·7764 $_{30}$	1·26	0·8962 $_{18}$	1·76	0·9608 $_{8}$	2·26	0·98809 $_{31}$
0·27	0·6064 $_{39}$	0·77	0·7794 $_{29}$	1·27	0·8980 $_{17}$	1·77	0·9616 $_{9}$	2·27	0·98840 $_{30}$
0·28	0·6103 $_{38}$	0·78	0·7823 $_{29}$	1·28	0·8997 $_{18}$	1·78	0·9625 $_{8}$	2·28	0·98870 $_{29}$
0·29	0·6141 $_{38}$	0·79	0·7852 $_{29}$	1·29	0·9015 $_{17}$	1·79	0·9633 $_{8}$	2·29	0·98899 $_{29}$
0·30	0·6179 $_{38}$	0·80	0·7881 $_{29}$	1·30	0·9032 $_{17}$	1·80	0·9641 $_{8}$	2·30	0·98928 $_{28}$
0·31	0·6217 $_{38}$	0·81	0·7910 $_{29}$	1·31	0·9049 $_{17}$	1·81	0·9649 $_{7}$	2·31	0·98956 $_{27}$
0·32	0·6255 $_{38}$	0·82	0·7939 $_{28}$	1·32	0·9066 $_{16}$	1·82	0·9656 $_{8}$	2·32	0·98983 $_{27}$
0·33	0·6293 $_{38}$	0·83	0·7967 $_{28}$	1·33	0·9082 $_{17}$	1·83	0·9664 $_{7}$	2·33	0·99010 $_{26}$
0·34	0·6331 $_{37}$	0·84	0·7995 $_{28}$	1·34	0·9099 $_{16}$	1·84	0·9671 $_{7}$	2·34	0·99036 $_{25}$

Table 4—continued

x	$\Phi(x)$	x	$\Phi(x)$	x	$\Phi(x)$	x	$\Phi(x)$	x	$\Phi(x)$
0·35	0·6368$_{38}$	0·85	0·8023$_{28}$	1·35	0·9115$_{16}$	1·85	0·9678$_{8}$	2·35	0·99061$_{25}$
0·36	0·6406$_{37}$	0·86	0·8051$_{27}$	1·36	0·9131$_{16}$	1·86	0·9686$_{7}$	2·36	0·99086$_{25}$
0·37	0·6443$_{37}$	0·87	0·8078$_{28}$	1·37	0·9147$_{15}$	1·87	0·9693$_{6}$	2·37	0·99111$_{23}$
0·38	0·6480$_{37}$	0·88	0·8106$_{27}$	1·38	0·9162$_{15}$	1·88	0·9699$_{7}$	2·38	0·99134$_{24}$
0·39	0·6517$_{37}$	0·89	0·8133$_{26}$	1·39	0·9177$_{15}$	1·89	0·9706$_{7}$	2·39	0·99158$_{22}$
0·40	0·6554$_{37}$	0·90	0·8159$_{27}$	1·40	0·9192$_{15}$	1·90	0·9713$_{6}$	2·40	0·99180$_{22}$
0·41.	0·6591$_{37}$	0·91	0·8186$_{26}$	1·41	0·9207$_{15}$	1·91	0·9719$_{7}$	2·41	0·99202$_{22}$
0·42	0·6628$_{36}$	0·92	0·8212$_{26}$	1·42	0·9222$_{14}$	1·92	0·9726$_{6}$	2·42	0·99224$_{21}$
0·43	0·6664$_{36}$	0·93	0·8238$_{26}$	1·43	0·9236$_{15}$	1·93	0·9732$_{6}$	2·43	0·99245$_{21}$
0·44	0·6700$_{36}$	0·94	0·8264$_{25}$	1·44	0·9251$_{14}$	1·94	0·9738$_{6}$	2·44	0·99266$_{20}$
0·45	0·6736$_{36}$	0·95	0·8289$_{26}$	1·45	0·9265$_{14}$	1·95	0·9744$_{6}$	2·45	0·99286$_{19}$
0·46	0·6772$_{36}$	0·96	0·8315$_{25}$	1·46	0·9279$_{13}$	1·96	0·9750$_{6}$	2·46	0·99305$_{19}$
0·47	0·6808$_{36}$	0·97	0·8340$_{25}$	1·47	0·9292$_{14}$	1·97	0·9756$_{5}$	2·47	0·99324$_{19}$
0·48	0·6844$_{35}$	0·98	0·8365$_{24}$	1·48	0·9306$_{13}$	1·98	0·9761$_{6}$	2·48	0·99343$_{18}$
0·49	0·6879$_{36}$	0·99	0·8389$_{24}$	1·49	0·9319$_{13}$	1·99	0·9767$_{5}$	2·49	0·99361$_{18}$
0·50	0·6915	1·00	0·8413	1·50	0·9332	2·00	0·9772	2·50	0·99379

x	$\Phi(x)$	x	$\Phi(x)$	x	$\Phi(x)$
2·50	0·99379$_{17}$	2·70	0·99653$_{11}$	2·90	0·99813$_{6}$
2·51	0·99396$_{17}$	2·71	0·99664$_{10}$	2·91	0·99819$_{6}$
2·52	0·99413$_{17}$	2·72	0·99674$_{9}$	2·92	0·99825$_{6}$
2·53	0·99430$_{16}$	2·73	0·99683$_{10}$	2·93	0·99831$_{5}$
2·54	0·99446$_{15}$	2·74	0·99693$_{9}$	2·94	0·99836$_{5}$
2·55	0·99461$_{16}$	2·75	0·99702$_{9}$	2·95	0·99841$_{5}$
2·56	0·99477$_{15}$	2·76	0·99711$_{9}$	2·96	0·99846$_{5}$
2·57	0·99492$_{14}$	2·77	0·99720$_{8}$	2·97	0·99851$_{5}$
2·58	0·99506$_{14}$	2·78	0·99728$_{8}$	2·98	0·99856$_{5}$
2·59	0·99520$_{14}$	2·79	0·99736$_{8}$	2·99	0·99861$_{4}$
2·60	0·99534$_{13}$	2·80	0·99744$_{8}$	3·0	0·99865$_{38}$
2·61	0·99547$_{13}$	2·81	0·99752$_{8}$	3·1	0·99903$_{28}$
2·62	0·99560$_{13}$	2·82	0·99760$_{7}$	3·2	0·99931$_{21}$
2·63	0·99573$_{12}$	2·83	0·99767$_{7}$	3·3	0·99952$_{14}$
2·64	0·99585$_{13}$	2·84	0·99774$_{7}$	3·4	0·99966$_{11}$

Table 4—continued

x	$\Phi(x)$	x	$\Phi(x)$	x	$\Phi(x)$
2·65	0·99598 $_{11}$	2·85	0·99781 $_7$	3·5	0·99977 $_7$
2·66	0·99609 $_{12}$	2·86	0·99788 $_7$	3·6	0·99984 $_5$
2·67	0·99621 $_{11}$	2·87	0·99795 $_6$	3·7	0·99989 $_4$
2·68	0·99632 $_{11}$	2·88	0·99801 $_6$	3·8	0·99993 $_2$
2·69	0·99643 $_{10}$	2·89	0·99807 $_6$	3·9	0·99995 $_2$
2·70	0·99653	2·90	0·99813	4·0	0·99997

The function tabulated is $\Phi(x) = \dfrac{1}{\sqrt{(2\pi)}} \displaystyle\int_{-\infty}^{\infty} \exp(-\tfrac{1}{2}t^2)\, dt$. $\Phi(x)$ is the probability that a random variable normally distributed with zero mean and unit variance, will be less than x. The last two columns give the ordinate $\phi(x) = \dfrac{1}{\sqrt{(2\pi)}} e^{-\frac{1}{2}x^2}$ of the normal frequency curve.

Table 5. Percentage Points of the t-Distribution

P	25	10	5	2	1	0·2	0·1	$\dfrac{120}{\nu}$
$\nu = 1$	2·41	6·31	12·71	31·82	63·66	318·3	636·6	
2	1·60	2·92	4·30	6·96	9·92	22·33	31·60	
3	1·42	2·35	3·18	4·54	5·84	10·21	12·92	
4	1·34	2·13	2·78	3·75	4·60	7·17	8·61	
5	1·30	2·02	2·57	3·36	4·03	5·89	6·87	
6	1·27	1·94	2·45	3·14	3·71	5·21	5·96	
7	1·25	1·89	2·36	3·00	3·50	4·79	5·41	
8	1·24	1·86	2·31	2·90	3·36	4·50	5·04	
9	1·23	1·83	2·26	2·82	3·25	4·30	4·78	
10	1·22	1·81	2·23	2·76	3·17	4·14	4·59	12
12	1·21	1·78	2·18	2·68	3·05	3·93	4·32	10
15	1·20	1·75	2·13	2·60	2·95	3·73	4·07	8
20	1·18	1·72	2·09	2·53	2·85	3·55	3·85	6
24	1·18	1·71	2·06	2·49	2·80	3·47	3·75	5
30	1·17	1·70	2·04	2·46	2·75	3·39	3·65	4
40	1·17	1·68	2·02	2·42	2·70	3·31	3·55	3
60	1·16	1·67	2·00	2·39	2·66	3·23	3·46	2
120	1·16	1·66	1·98	2·36	2·62	3·16	3·37	1
∞	1·15	1·64	1·96	2·33	2·58	3·09	3·29	0

The function tabulated is t_P defined by the equation

$$\frac{P}{100} = \frac{1}{\sqrt{(v\pi)}} \frac{\Gamma(\tfrac{1}{2}v + \tfrac{1}{2})}{\Gamma(\tfrac{1}{2}v)} \int_{|t| \geqslant t_P} \frac{dt}{(1 + t^2/v)^{\frac{1}{2}(v+1)}}.$$

If t is the ratio of a random variable, normally distributed with zero mean, to an independent estimate of its standard deviation based on v degrees of freedom, $P/100$ is the probability that $|t| \geqslant t_P$.

Interpolation v-wise should be linear in $120/v$.

Other percentage points may be found approximately, except when v and P are both small, by using the fact that the variable

$$y = \pm \sinh^{-1} \sqrt{(3t^2/2v)},$$

where y has the same sign as t, is approximately normally distributed with zero mean and variance $3/(2v - 1)$.

Table 6. Percentage Points of the χ^2-Distribution

P	99·5	99	97·5	95	10	5	2·5	1	0·5	0·1
v = 1	0·0⁴393	0·0³157	0·0³982	0·00393	2·71	3·84	5·02	6·63	7·88	10·83
2	0·0100	0·0201	0·0506	0·103	4·61	5·99	7·38	9·21	10·60	13·81
3	0·0717	0·115	0·216	0·352	6·25	7·81	9·35	11·34	12·84	16·27
4	0·207	0·297	0·484	0·711	7·78	9·49	11·14	13·28	14·86	18·47
5	0·412	0·554	0·831	1·15	9·24	11·07	12·83	15·09	16·75	20·52
6	0·676	0·872	1·24	1·64	10·64	12·59	14·45	16·81	18·55	22·46
7	0·989	1·24	1·69	2·17	12·02	14·07	16·01	18·48	20·28	24·32
8	1·34	1·65	2·18	2·73	13·36	15·51	17·53	20·09	21·95	26·12
9	1·73	2·09	2·70	3·33	14·68	16·92	19·02	21·67	23·59	27·88
10	2·16	2·56	3·25	3·94	15·99	18·31	20·48	23·21	25·19	29·59
11	2·60	3·05	3·82	4·57	17·28	19·68	21·92	24·73	26·76	31·26
12	3·07	3·57	4·40	5·23	18·55	21·03	23·34	26·22	28·30	32·91
13	3·57	4·11	5·01	5·89	19·81	22·36	24·74	27·69	29·82	34·53
14	4·07	4·66	5·63	6·57	21·06	23·68	26·12	29·14	31·32	36·12
15	4·60	5·23	6·26	7·26	22·31	25·00	27·49	30·58	32·80	37·70
16	5·14	5·81	6·91	7·96	23·54	26·30	28·85	32·00	34·27	39·25
17	5·70	6·41	7·56	8·67	24·77	27·59	30·19	33·41	35·72	40·79
18	6·26	7·01	8·23	9·39	25·99	28·87	31·53	34·81	37·16	42·31
19	6·84	7·63	8·91	10·12	27·20	30·14	32·85	36·19	38·58	43·82
20	7·43	8·26	9·59	10·85	28·41	31·41	34·17	37·57	40·00	45·31
21	8·03	8·90	10·28	11·59	29·62	32·67	35·48	38·93	41·40	46·80
22	8·64	9·54	10·98	12·34	30·81	33·92	36·78	40·29	42·80	48·27
23	9·26	10·20	11·69	13·09	32·01	35·17	38·08	41·64	44·18	49·73
24	9·89	10·86	12·40	13·85	33·20	36·42	39·36	42·98	45·56	51·18
25	10·52	11·52	13·12	14·61	34·38	37·65	40·65	44·31	46·93	52·62
26	11·16	12·20	13·84	15·38	35·56	38·89	41·92	45·64	48·29	54·05
27	11·81	12·88	14·57	16·15	36·74	40·11	43·19	46·96	49·64	55·48
28	12·46	13·56	15·31	16·93	37·92	41·34	44·46	48·28	50·99	56·89
29	13·12	14·26	16·05	17·71	39·09	42·56	45·72	49·59	52·34	58·30

Table 6—continued

P	99·5	99	97·5	95	10	5	2·5	1	0·5	0·1
30	13·79	14·95	16·79	18·49	40·26	43·77	46·98	50·89	53·67	59·70
40	20·71	22·16	24·43	26·51	51·81	55·76	59·34	63·69	66·77	73·40
50	27·99	29·71	32·36	34·76	63·17	67·50	71·42	76·15	79·49	86·66
60	35·53	37·48	40·48	43·19	74·40	79·08	83·30	88·38	91·95	99·61
70	43·28	45·44	48·76	51·74	85·53	90·53	95·02	100·4	104·2	112·3
80	51·17	53·54	57·15	60·39	96·58	101·9	106·6	112·3	116·3	124·8
90	59·20	61·75	65·65	69·13	107·6	113·1	118·1	124·1	128·3	137·2
100	67·33	70·06	74·22	77·93	118·5	124·3	129·6	135·8	140·2	149·4

The function tabulated is χ_P^2 defined by the equation $\dfrac{P}{100} = \dfrac{1}{2^{v/2}\Gamma(\frac{1}{2}v)} \displaystyle\int_{\chi_P^2}^{\infty} x^{\frac{1}{2}v - 1} e^{-x/2} \, dx$. If x is a variable distributed as χ^2 with v degrees of freedom, $P/100$ is the probability that $x \geqslant \chi_P^2$. For $v < 100$, linear interpolation in v is adequate. For $v > 100$, $\sqrt{(2\chi^2)}$ is approximately normally distributed with mean $\sqrt{(2v - 1)}$ and unit variance.

Table 7(a). 5 Per Cent Points of the F-Distribution

$v_1 =$	1	2	3	4	5	6	7	8	10	12	24	∞
$v_2 = 1$	161·4	199·5	215·7	224·6	230·2	234·0	236·8	238·9	241·9	243·9	249·0	254·3
2	18·5	19·0	19·2	19·2	19·3	19·3	19·4	19·4	19·4	19·4	19·5	19·5
3	10·13	9·55	9·28	9·12	9·01	8·94	8·89	8·85	8·79	8·74	8·64	8·53
4	7·71	6·94	6·59	6·39	6·26	6·16	6·09	6·04	5·96	5·91	5·77	5·63
5	6·61	5·79	5·41	5·19	5·05	4·95	4·88	4·82	4·74	4·68	4·53	4·36
6	5·99	5·14	4·76	4·53	4·39	4·28	4·21	4·15	4·06	4·00	3·84	3·67
7	5·59	4·74	4·35	4·12	3·97	3·87	3·79	3·73	3·64	3·57	3·41	3·23
8	5·32	4·46	4·07	3·84	3·69	3·58	3·50	3·44	3·35	3·28	3·12	2·93
9	5·12	4·26	3·86	3·63	3·48	3·37	3·29	3·23	3·14	3·07	2·90	2·71
10	4·96	4·10	3·71	3·48	3·33	3·22	3·14	3·07	2·98	2·91	2·74	2·54
11	4·84	3·98	3·59	3·36	3·20	3·09	3·01	2·95	2·85	2·79	2·61	2·40
12	4·75	3·89	3·49	3·26	3·11	3·00	2·91	2·85	2·75	2·69	2·51	2·30
13	4·67	3·81	3·41	3·18	3·03	2·92	2·83	2·77	2·67	2·60	2·42	2·21
14	4·60	3·74	3·34	3·11	2·96	2·85	2·76	2·70	2·60	2·53	2·35	2·13
15	4·54	3·68	3·29	3·06	2·90	2·79	2·71	2·64	2·54	2·48	2·29	2·07
16	4·49	3·63	3·24	3·01	2·85	2·74	2·66	2·59	2·49	2·42	2·24	2·01
17	4·45	3·59	3·20	2·96	2·81	2·70	2·61	2·55	2·45	2·38	2·19	1·96
18	4·41	3·55	3·16	2·93	2·77	2·66	2·58	2·51	2·41	2·34	2·15	1·92
19	4·38	3·52	3·13	2·90	2·74	2·63	2·54	2·48	2·38	2·31	2·11	1·88
20	4·35	3·49	3·10	2·87	2·71	2·60	2·51	2·45	2·35	2·28	2·08	1·84
21	4·32	3·47	3·07	2·84	2·68	2·57	2·49	2·42	2·32	2·25	2·05	1·81
22	4·30	3·44	3·05	2·82	2·66	2·55	2·46	2·40	2·30	2·23	2·03	1·78
23	4·28	3·42	3·03	2·80	2·64	2·53	2·44	2·37	2·27	2·20	2·00	1·76
24	4·26	3·40	3·01	2·78	2·62	2·51	2·42	2·36	2·25	2·18	1·98	1·73

APPENDIX

Table 7a—continued

$v_1 =$	1	2	3	4	5	6	7	8	10	12	24	∞
25	4·24	3·39	2·99	2·76	2·60	2·49	2·40	2·34	2·24	2·16	1·96	1·71
26	4·23	3·37	2·98	2·74	2·59	2·47	2·39	2·32	2·22	2·15	1·95	1·69
27	4·21	3·35	2·96	2·73	2·57	2·46	2·37	2·31	2·20	2·13	1·93	1·67
28	4·20	3·34	2·95	2·71	2·56	2·45	2·36	2·29	2·19	2·12	1·91	1·65
29	4·18	3·33	2·93	2·70	2·55	2·43	2·35	2·28	2·18	2·10	1·90	1·64
30	4·17	3·32	2·92	2·69	2·53	2·42	2·33	2·27	2·16	2·09	1·89	1·62
32	4·15	3·29	2·90	2·67	2·51	2·40	2·31	2·24	2·14	2·07	1·86	1·59
34	4·13	3·28	2·88	2·65	2·49	2·38	2·29	2·23	2·12	2·05	1·84	1·57
36	4·11	3·26	2·87	2·63	2·48	2·36	2·28	2·21	2·11	2·03	1·82	1·55
38	4·10	3·24	2·85	2·62	2·46	2·35	2·26	2·19	2·09	2·02	1·81	1·53
40	4·08	3·23	2·84	2·61	2·45	2·34	2·25	2·18	2·08	2·00	1·79	1·51
60	4·00	3·15	2·76	2·53	2·37	2·25	2·17	2·10	1·99	1·92	1·70	1·39
120	3·92	3·07	2·68	2·45	2·29	2·18	2·09	2·02	1·91	1·83	1·61	1·25
∞	3·84	3·00	2·60	2·37	2·21	2·10	2·01	1·94	1·83	1·75	1·52	1·00

The above table gives values of $F_{5\%}$ where $\dfrac{5}{100}$ is the probability of obtaining a value of $F > F_{5\%}$ for various combinations of v_1 and v_2.

Table 7(b). 1 Per Cent Points of the F-Distribution

$v_1 =$	1	2	3	4	5	6	7	8	10	12	24	∞
$v_2 = 1$	4052	5000	5403	5625	5764	5859	5928	5981	6056	6106	6235	6366
2	98·5	99·0	99·2	99·2	99·3	99·3	99·4	99·4	99·4	99·4	99·5	99·5
3	34·1	30·8	29·5	28·7	28·2	27·9	27·7	27·5	27·2	27·1	26·6	26·1
4	21·2	18·0	16·7	16·0	15·5	15·2	15·0	14·8	14·5	14·4	13·9	13·5
5	16·26	13·27	12·06	11·39	10·97	10·67	10·46	10·29	10·05	9·89	9·47	9·02
6	13·74	10·92	9·78	9·15	8·75	8·47	8·26	8·10	7·87	7·72	7·31	6·88
7	12·25	9·55	8·45	7·85	7·46	7·19	6·99	6·84	6·62	6·47	6·07	5·65
8	11·26	8·65	7·59	7·01	6·63	6·37	6·18	6·03	5·81	5·67	5·28	4·86
9	10·56	8·02	6·99	6·42	6·06	5·80	5·61	5·47	5·26	5·11	4·73	4·31
10	10·04	7·56	6·55	5·99	5·64	5·39	5·20	5·06	4·85	4·71	4·33	3·91
11	9·65	7·21	6·22	5·67	5·32	5·07	4·89	4·74	4·54	4·40	4·02	3·60
12	9·33	6·93	5·95	5·41	5·06	4·82	4·64	4·50	4·30	4·16	3·78	3·36
13	9·07	6·70	5·74	5·21	4·86	4·62	4·44	4·30	4·10	3·96	3·59	3·17
14	8·86	6·51	5·56	5·04	4·70	4·46	4·28	4·14	3·94	3·80	3·43	3·00
15	8·68	6·36	5·42	4·89	4·56	4·32	4·14	4·00	3·80	3·67	3·29	2·87
16	8·53	6·23	5·29	4·77	4·44	4·20	4·03	3·89	3·69	3·55	3·18	2·75
17	8·40	6·11	5·18	4·67	4·34	4·10	3·93	3·79	3·59	3·46	3·08	2·65
18	8·29	6·01	5·09	4·58	4·25	4·01	3·84	3·71	3·51	3·37	3·00	2·57
19	8·18	5·93	5·01	4·50	4·17	3·94	3·77	3·63	3·43	3·30	2·92	2·49

Table 7b—continued

$v_1 =$	1	2	3	4	5	6	7	8	10	12	24	∞
20	8·10	5·85	4·94	4·43	4·10	3·87	3·70	3·56	3·37	3·23	2·86	2·42
21	8·02	5·78	4·87	4·37	4·04	3·81	3·64	3·51	3·31	3·17	2·80	2·36
22	7·95	5·72	4·82	4·31	3·99	3·76	3·59	3·45	3·26	3·12	2·75	2·31
23	7·88	5·66	4·76	4·26	3·94	3·71	3·54	3·41	3·21	3·07	2·70	2·26
24	7·82	5·61	4·72	4·22	3·90	3·67	3·50	3·36	3·17	3·03	2·66	2·21
25	7·77	5·57	4·68	4·18	3·86	3·63	3·46	3·32	3·13	2·99	2·62	2·17
26	7·72	5·53	4·64	4·14	3·82	3·59	3·42	3·29	3·09	2·96	2·58	2·13
27	7·68	5·49	4·60	4·11	3·78	3·56	3·39	3·26	3·06	2·93	2·55	2·10
28	7·64	5·45	4·57	4·07	3·75	3·53	3·36	3·23	3·03	2·90	2·52	2·06
29	7·60	5·42	4·54	4·04	3·73	3·50	3·33	3·20	3·00	2·87	2·49	2·03
30	7·56	5·39	4·51	4·02	3·70	3·47	3·30	3·17	2·98	2·84	2·47	2·01
32	7·50	5·34	4·46	3·97	3·65	3·43	3·26	3·13	2·93	2·80	2·42	1·96
34	7·45	5·29	4·42	3·93	3·61	3·39	3·22	3·09	2·90	2·76	2·38	1·91
36	7·40	5·25	4·38	3·89	3·58	3·35	3·18	3·05	2·86	2·72	2·35	1·87
38	7·35	5·21	4·34	3·86	3·54	3·32	3·15	3·02	2·83	2·69	2·32	1·84
40	7·31	5·18	4·31	3·83	3·51	3·29	3·12	2·99	2·80	2·66	2·29	1·80
60	7·08	4·98	4·13	3·65	3·34	3·12	2·95	2·82	2·63	2·50	2·12	1·60
120	6·85	4·79	3·95	3·48	3·17	2·96	2·79	2·66	2·47	2·34	1·95	1·38
∞	6·63	4·61	3·78	3·32	3·02	2·80	2·64	2·51	2·32	2·18	1·79	1·00

The above table gives values of $F_{1\%}$ where $\dfrac{1}{100}$ is the probability of obtaining a value of $F > F_{1\%}$ for various combinations of v_1 and v_2.

[*Tables 4, 5, 6, 7a* and *7b* are reproduced from Lindley, D. V. and Miller, J. C. P. (1966) *Cambridge Elementary Statistical Tables,* by courtesy of Cambridge University Press.

SOLUTIONS

Exercises 2b
   2. 8 classes.

Exercises 2c
   4. (*a*) Isotype representation of frequency polygon,
      (*b*) Bar chart or isotype representation,
      (*c*) Pie chart,
      (*d*) Bar chart.

Exercises 2d
   1. (*a*) Each number would be expected to occur 20 times,
      (*b*) 30 heads would be expected.

   2.

Score	2	3	4	5	6	7	8	9	10	11	12
Proportions	1	2	3	4	5	6	5	4	3	2	1
Expected	2	4	6	8	10	12	10	8	6	4	2

   3. It is a random sample.
   4. Yes, if there is no obstruction, e.g. traffic lights.

Exercises 2e
   5. Not a random sample.

EXERCISES 2

1. 16·045–16·095   5
   16·095–16·145   13
   16·145–16·195   22
   16·195–16·245   23
   16·245–16·295   13
   16·295–16·345   4
2. Total weight from the original data 1295·50 oz. Total weight from the frequency table 1295·50 oz.
3. A pie chart is not suitable because the data is not all the subdivisions of a single subject.
5. Using a bar chart is the most suitable method.
6. £5028·5.
7. Discrete (*a*) (*f*); continuous (*b*), (*c*), (*d*), (*e*), (*g*).

9. Although some people live to be over 100 years old, for numerical working we have to allocate the centre of the class interval to the frequency and 89 is a suitable choice for an upper limit to the last class.

10.

Age (years)	(months)	%
15	11	74·5
15	10	76·8
	9	84·7
	8	77·1
	7	76·8
	6	74·4
	5	78·3
	4	82·2
	3	81·3
	2	79·8
	1	63·0
	0	89·2

11. The bar chart is the better method for comparison purposes. More especially if different shading is used for the bars representing the 1948 and 1949 figures.

13. The numbers in two categories have been changed around; correct reading should be: 147 male employees, 312 coloured married.

14. 25 per cent.

15. The histogram is skew.

Exercises 3a

1. (b) and (c) are mutually exclusive.
2. (a) and (c) are independent.
3. (a)   (b)   (c)

(d)

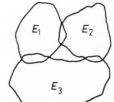

4. (a) $\dfrac{1}{12}$, (b) $\dfrac{1}{6}$, (c) 0, (d) 0, (e) $\dfrac{1}{2}$.

5. The theoretical probabilities for each event are:

(a) $\Pr(\text{H}) = \dfrac{1}{2} = \Pr(T)$

(b)

Score	2	3	4	5	6	7	8	9	10	11	12
Theoretical probability	$\dfrac{1}{36}$	$\dfrac{2}{36}$	$\dfrac{3}{36}$	$\dfrac{4}{36}$	$\dfrac{5}{36}$	$\dfrac{6}{36}$	$\dfrac{5}{36}$	$\dfrac{4}{36}$	$\dfrac{3}{36}$	$\dfrac{2}{36}$	$\dfrac{1}{36}$

(c) $\Pr(\text{Heart}) = \Pr(\text{Spade}) = \Pr(\text{Club}) = \Pr(\text{Diamond}) = \dfrac{1}{4}$.

6. (a) $\dfrac{1}{8}$, (b) $\dfrac{7}{8}$.

7. Expected $\Pr(3 \text{ Heads}) = \dfrac{1}{4}$.

9. (a) $\dfrac{2}{5}$, (b) $\dfrac{3}{5}$, (c) $\dfrac{3}{25}$, (d) $\dfrac{4}{25}$.

10. A's chance is $\dfrac{2}{3}$, B's chance is $\dfrac{1}{3}$.

Exercises 3b

1. 24. 2. $3^4 = 81$. 3. 12 ways.

4. 6 ways. In question 3 the order was important, therefore AB is a different way of voting than BA. In this question the order is *not* important, the vote AB or BA returns the same candidates.

5. $6 \times 5 \times 4 = 120$ ways.

Exercises 3c

1. (a) 20, 120, 5040, 360
   (b) $3 \cdot 847 \times 10^{20}$
       $7 \cdot 281 \times 10^{21}$
       $7 \cdot 268 \times 10^{13}$
       $2 \cdot 746 \times 10^{150}$
   (c) $(n + 1)!$
       $\dfrac{9!}{2^4 \times 4!}$

2. 3,628,800; £45,360.   3. 1680.
4. 120.   5. 40,320.

Exercises 3d

1. (a) 3003, 2024, 126;   (b) $n = 16$.

2. (a) 330, (b) 35, (c) 295.

3. (a) 1326, (b) 6, (c) $\dfrac{1}{221}$ 4. 2520.

5. (a) 142,506, (b) 42,504, (c) 136,620.

Exercises 3e

1.
Score	3	4	5	6	7	8	9	10	11	12	13	14	15	16	17	18
Probability	$\frac{1}{216}$	$\frac{3}{216}$	$\frac{6}{216}$	$\frac{10}{216}$	$\frac{15}{216}$	$\frac{21}{216}$	$\frac{25}{216}$	$\frac{27}{216}$	$\frac{27}{216}$	$\frac{25}{216}$	$\frac{21}{216}$	$\frac{15}{216}$	$\frac{10}{216}$	$\frac{6}{216}$	$\frac{3}{216}$	$\frac{1}{216}$

2. $\Pr(x) = \dfrac{(x-1)(x-2)}{2}$ $(3 \leqslant x \leqslant 8)$

Alternatively the sequence 1, 3, 6, 10, ... is obtained by adding successively 2, 3, 4, ... If the law held then Pr(9) = 28, and Pr(10) = 36. The discrepancy arises because 6 is the highest number available on the dice and the formula value includes the cases 7, 1, 1 ; 1, 7, 1 ; 1, 1, 7 which are impossible in practice.

3.
Score	2	3	4	5	6	7	8	9	10
Probability	0·01	0·02	0·03	0·04	0·05	0·06	0·07	0·08	0·09

11	12	13	14	15	16	17	18	19	20
0·10	0·09	0·08	0·07	0·06	0·05	0·04	0·03	0·02	0·01

4.
No. of heads	5	4	3	2	1	0
Probability	$\frac{1}{32}$	$\frac{5}{32}$	$\frac{10}{32}$	$\frac{10}{32}$	$\frac{5}{32}$	$\frac{1}{32}$

Exercises 3f

1.
x	1	2	3	4	5	6
p	$\frac{1}{6}$	$\frac{1}{6}$	$\frac{1}{6}$	$\frac{1}{6}$	$\frac{1}{6}$	$\frac{1}{6}$

$\mathscr{E}(x) = 3\cdot5$

2.
H	0	1	2	3
p	$\frac{1}{8}$	$\frac{3}{8}$	$\frac{3}{8}$	$\frac{1}{8}$

$\mathscr{E}(H) = 1\cdot5$

3.

H	0	1	2	3	4
p	$\dfrac{1}{16}$	$\dfrac{4}{16}$	$\dfrac{6}{16}$	$\dfrac{4}{16}$	$\dfrac{1}{16}$

$\mathscr{E}(H) = 2$

4. *No*, because the 60 throws are only a sample of all possible throws and the sample mean is not, in general, equal to the population mean.

5. *No* for the same reason as in question 4.

EXERCISES 3

1. (a) 216, (b) 6, (c) $\dfrac{1}{36}$.

2. $\dfrac{1}{12}$, yes.

3. 2,598,960.

4. (a) $n = 6$, (b) (i) 0·21, (ii) 0·20

5. $\dfrac{29}{30}$.

6. 362,880, 120.

7. (a) $\dfrac{19}{72}$, (b) $\dfrac{53}{72}$.

8. (a) $p_1 p_2$, (b) $p_1(1 - p_2)$,
   (c) $(1 - p_1)p_2$,
   (d) $(1 - p_1)(1 - p_2)$

9. (a) 3136, (b) 2296, (c) $\dfrac{3}{8}$.

10. (a) (i) $\dfrac{1}{22}$, (ii) $\dfrac{3}{11}$;

   (b) (i) 0·230, (ii) 0·211.

11. $1 - \left(\dfrac{39}{40}\right)^5, 1 - \left(\dfrac{39}{40}\right)^{10}, 1 - \left(\dfrac{39}{40}\right)^{50}, 1 - \left(\dfrac{39}{40}\right)^n$.

12. (a) 5148, (b) 111,540, (c) 36, (d) 9216.

13. $\frac{1}{2}$.

14. (a) £1, (b) Loss of £100.

15. (a) $\dfrac{1}{285}$, (b) $\dfrac{1}{57}$, (c) $\dfrac{32}{57}$.

16. 300, 1080.

17. B's chance is either $1 -$ A's, i.e. $1 - p$, or after A has lost on the first try it will then be p, hence taking into account the probability of A's losing it will then be $\dfrac{5}{6} p$, which equals $1 - p$. $\therefore p = \dfrac{6}{11}$.

18. $\dfrac{2}{15}$.

19. $\dfrac{11}{24}$.

20. $\dfrac{9}{220}$, $\dfrac{7}{12}$.

21. $\mathscr{E}(x) = 3·5$.

22. $\dfrac{2}{(n - 1)}$.

23. (a) 38,760, (b) 18,740.

24. $1 - \left(\dfrac{19}{20}\right)^{10}$.

25. (a) 0·0087, (b) 0·33, (c) 0·71.

26. (a) 0·4512, (b) $\dfrac{b!\,w!}{(b+w)!}$.

27. (a) (i) $\dfrac{9}{91}$, (ii) $\dfrac{15}{91}$;

(b) (i) $\dfrac{2793}{3125}$, (ii) $\dfrac{2563}{25000}$

28. (a) $\dfrac{37}{108}$, (b) $\dfrac{1}{15}, \dfrac{1}{15}, \dfrac{2}{15}, \dfrac{2}{15}, \dfrac{3}{15}, \dfrac{2}{15}, \dfrac{2}{15}, \dfrac{1}{15}, \dfrac{1}{15}$, (c) 0·504.

29. (a) 0·35, $n = 44$; (b) (i) $\dfrac{1}{5}$, (ii) $\dfrac{1}{5}$.

Exercises 4a

1. (a) $\dfrac{1}{2^6}(1 + 6 + 15 + 20 + 15 + 6 + 1)$, (b) $\dfrac{1}{4^5}(1 + 15 + 90 + 270 + 405 + 243)$, (c) $1 + 12 + 60 + 160 + 240 + 192 + 64$, (d) $1 + 20 + 160 + 640 + 1280 + 1024$.

2. (a) $\dfrac{1}{3^6}(1, 12, 60, 160, 240, 192, 64)$, (b) 160.

3. (a) $\dfrac{1}{4^5}$, (b) $1 - \dfrac{243}{4^5}$.

4. (a) $\dfrac{288}{78,125}$, (b) $\dfrac{3024}{15,625}$.

5. $\dfrac{211}{16,384}$.

6. $p = 0·1$; 59, 32·8, 7·3, 0·8, 0·1, 0·0

Exercises 4b

1. 0·0652.

2. 0·0189, 0·0529.

3

x	Binomial	Poisson
0	0·363	0·368
1	0·373	0·368
2	0·187	0·184
3	0·060	0·061
4	0·014	0·015
5	0·003	0·003

4. 0·28 per cent.

5. Mean is 1·9; expected frequencies: 15·0, 28·4, 27·0, 17·1, 8·1, 3·1.

EXERCISES 4

1. (a) $\dfrac{63}{64}$, (b) $\dfrac{31}{32}$, (c) $\dfrac{5}{16}$.

2. (a) $\dfrac{256}{625}$, (b) $\dfrac{32}{625}$, (c) $\dfrac{3104}{3125}$.

3. (a) 0·0067, 0·0337, 0·0842, 0·1404;
   (b) 0·6065, 0·3033, 0·0758, 0·0126;

(c) 0·0408, 0·1304, 0·2087, 0·2226.

4. (a) 0·273, (b) 0·450, (c) 0·373.

5. (a) The hits are at random,
   (b) mean is 1·40,
   (c) 24·7, 34·5, 24·2, 11·3, 3·9, 1·1.

6. (a) $\dfrac{1}{10,000}$, (b) $\dfrac{9999}{10,000}$, (c) $\dfrac{9477}{10,000}$.

7. (a) 0·423, (b) 0·084.

8. 0·9197, 0·2424. 9. 0·368, 0·368, 0·184, 0·061, 0·015.

10. 0·242. 11. (a) 0·056, (b) 0·061.

12. 7·62.

	0	1	2	3
(a)	0·004	0·031	0·109	0·219
(b)	0·012	0·058	0·137	0·205
(c)	0·015	0·065	0·142	0·206
(d)	0·018	0·073	0·146	0·195

14. (a) $^{20}C_{15}\left(\dfrac{3}{5}\right)^{15}\left(\dfrac{2}{5}\right)^{5} = 0.094$

(b) $^{28}C_{10}\left(\dfrac{3}{5}\right)^{10}\left(\dfrac{2}{5}\right)^{10} + {}^{20}C_{15}\left(\dfrac{3}{5}\right)^{15}\left(\dfrac{2}{5}\right)^{5} + \left(\dfrac{3}{5}\right)^{20} = 0.211.$

15. 1·43. 16. 1·62, 3·8, 4·5, 3·5, 2·1, 1·55.

17. Mean is 3·87. Expected frequencies 54·5, 211, 408·2, 526·8, 509·6, 394·1, 254·3, 143·4, 69·4, 29·7, 11·4, 3·9, 1·3.

18. 0·46, 0·46.

19. 0·185, 0·427; add another telephone line.

20. (a) 0·191, (b) 0·0837.

21. Mean 0·26, Pr(faulty blade) = 0·052.

Exercises 5a

1. (a) $\bar{x} = 4.213$, (b) $\bar{x} = 50.25$. 2. $\bar{x} = 8.32$.

3. $\bar{x}_A = 35.3$, $\bar{x}_B = 37.6$. 4. $\bar{x}_A = 4.949$, $\bar{x}_B = 5.305$.

5. The sample mean is 2·5126, thus there is no reason to reset the machine.

6. $\bar{x} = 15.92$.

Exercises 5b

1. Median 632·9, mode 634·03.

2. (a) median 612·95, mode 614·03;
   (b) median 1225·9, mode 1228·06;
   (c) median 421·84, mode 422·68.

3. Median 627·05, mode 525·97.   4. Median 4·227, mode 4·244.
5. Median 46·2, 25 percentile 39·4, 75 percentile 57·9.

EXERCISES 5

1. Mean 37·54, median 35·77.   2. 99 Ω.

3. $\dfrac{f_1\bar{x}_1 + f_2\bar{x}_2}{f_1 + f_2}$.   4. Median 35·13, mean 36·4.

5. (a) zero, (b) we deduce that 27·5 is the mean.
6. Mean 8·66, mode 9·0, median 9·0.
7. 46·08; 46·92.   8. 4·83, 4·72.
9. 6·912.   10. $2a - 3, 2b - 3, 2c - 3$.
11. 38·9 mile/h.   12. Mean 27·5, mode 32·7,
    median 29·0.
13. 22·4.

Exercises 6a

1. $\bar{x}_A = 16·06$, $\bar{x}_B = 16·06$; mean deviations 0·014 oz, 0·042 oz.
2. $\mu = 15$, mean deviation 3·55˙ (a) 3·55˙, (b) 3·55˙, (c) 17·77˙, (d) 1·77˙.
3. Median 50·0 Ω, mean 50·0 Ω, mean deviation 1·67 Ω.

4.

	Textiles	Food
Median	110·5	108·5
Mean deviation	9·1	4·8

Exercises 6b

1. $\sigma_A = 0·018$, $\sigma_B = 0·047$ 2. $\dfrac{1}{12}(n^2 - 1)$.

3. (a) σ, (b) σ, (c) not known, (d) $\dfrac{\sigma}{2}$

4 The climates have a different variation.   5. 9·9 in., 405 lb.
   Var(height) = 10·03, var(weight) = 4·136.

Exercises 6c

1.

	Height	Weight
Mean	67·3	163 lb
Coefficient variation	0·15	2·5 lb

2. (a) 0·0207, (b) 0·0220.

EXERCISES 6

1. 338 h.
3. Mean 2·001, standard deviation 0·031.
4.

	(a	(b)
Mode	6·0	3·15
Median	6·0	3·24
Mean	5·6	3·28
Range	9·0	0·80
Mean deviation about the mean	1·59	1·48
Standard deviation	2·02	0·18

5. Pearson's measure of skewness: -0.19, 0.72; coefficient variation 0.36, 0.55.
6. Mean $= 10.1$, variance $= 53.7$.   7. $N = 20$ or 25.
9. $\bar{x} = 4.596$, standard deviation $= 5.005$, 75 percentile $= 6.67$.
11. $\bar{x} = 1.0586$, $\mu_3(\bar{x}) = -0.0037$, $\mu_4(\bar{x}) = 0.0005$.
12. $\text{Var} = \dfrac{50^2}{400} + \dfrac{50^2}{600}$, $z \fallingdotseq 2$, significant at the 5 per cent level.
13. (a) $a \times$ mean $+ b$, (b) $a \times$ median $+ b$, (c) $a \times$ upper quartile $+ b$, (d) $a^2 \times$ variance, (e) $a \times$ standard deviation, (f) $a \times$ mean deviation, (g) $a \times$ semi-interquartile range.
14. 67·8, 7·06.
15. Mean 7 min 40·1 sec, standard deviation 17·3 sec, (a) 7 min 55 sec, (b) 63 percentile.
16. $m = 65.35$, $s = 16.90$, $\alpha = 25.87$.

Exercises 7a

1. $k = \dfrac{6}{125}$.

Exercises 7b

1. Variance $= 1$.   2. (a) $x = 1$, (b) $x = 2$.
3. (a) 1, (b) 1·25, (c) 2.
4. (a) mean 1, mean deviation $\dfrac{6}{e} - 1$, (b) mean $\dfrac{a}{2}$, mean deviation $\dfrac{a}{4}$.
5. $k = 0.75$, mean 1, variance 0·2.   6. 0·707.

EXERCISES 7

1. $k = \dfrac{3}{176}$.   2. $k = \dfrac{1}{6}$.

4. $\mathscr{E}(x) = 0.75$, $\mathscr{E}(x^2) = 0.6$.   5 $\dfrac{9}{16}$.

6. Mean 10, variance $56\frac{1}{7}$. 7. Mean $\frac{7\pi}{3}$, variance $\frac{34\pi^2}{45}$.

8. Mode $\frac{1}{2}$, mean $\frac{7}{6}$.

9. $k = \frac{3e^3}{e^3 - 1} = 3 \cdot 157$, mean $0 \cdot 281$.

10. $k = 1$, mean 2.

11. $k = \frac{15}{1024}$, mean $\frac{16}{7}$, variance $\frac{24}{49}$.

12. Mean $12\frac{1}{3}$, variance $\frac{184}{45}$. 13. $\frac{4}{9}$.

14. (a) $\frac{1}{a^2}$, (b) $2a$, (c) $6a^2$.

15. Pr(road damage) $= \frac{16}{25}$ (one bomb), probability with two bombs $\frac{544}{625}$.

16. $k = \frac{1}{\theta(e - 1)}$, $\mathscr{E}(x) = \frac{\theta(e - 2)}{(e - 1)}$. 17. $A = 2 \cdot 314$, mean $= 0 \cdot 343$, variance $= 0 \cdot 069$.

18. (a) mean $= 0$, standard deviation $= \sqrt{3}$; (b) $\frac{4a}{\pi} \cos \theta$.

Exercises 8a
1. (a) $0 \cdot 99693$, (b) $0 \cdot 9990$, (c) $0 \cdot 7611$.
2. (a) $0 \cdot 0418$, (b) $0 \cdot 00402$, (c) $0 \cdot 00097$.
3. (a) $0 \cdot 0548$, (b) $0 \cdot 00023$, (c) $0 \cdot 00554$.
4. (a) $0 \cdot 97778$, (b) $0 \cdot 9192$, (c) $0 \cdot 99993$.
5. (a) $0 \cdot 1632$, (b) $0 \cdot 07309$, (c) $0 \cdot 3432$.
6. (a) $0 \cdot 5085$, (b) $0 \cdot 7172$, (c) $0 \cdot 4222$.
7. (a) $0 \cdot 1298$, (b) $0 \cdot 07309$, (c) $0 \cdot 2421$.

Exercises 8b
1. $\mu = 33$, $\sigma = 10$.

EXERCISES 8

1. (a) $0 \cdot 8974$, (b) $0 \cdot 0786$, (c) $0 \cdot 3151$, (d) $0 \cdot 0495$, (e) $0 \cdot 0197$, (f) $0 \cdot 0154$, (g) $0 \cdot 6218$, (h) $0 \cdot 0541$.

2. (a) 0·0274, (b) 0·95, (c) 0·1587, (d) 0·50.

3. (a) 0·75, (b) −2·02, (c) 1·93, (d) $z = 1·07$ or 0·081.

4. (a) 17 lb, (b) 143–181 lb. 5. (a) 0·016, (b) 0·0453.

6. 75 per cent. 7. 8, 51, 133, 140, 58, 10.

8. 40. 9. 730.

10. (a) 109 h, (b) 0·6496, (c) 2075–2325 h.

11. Order of merit a, b, c. 12. 0·4602.

13. 36. 14. (a) 0·0898, (b) 0·09.

15. (a) 0·0558, (b) 0·0142. 16. (a) 0·109, (b) 0·0095, (c) 0·0413.

17. 0·0079. 18. 0·0083.

19. 0·0533. 20. 0·01368.

21. 33. 22. (a) 0·0276, (b) 0·2358, (c) 0·747.

23. $\bar{x} = 1002·45, s = 10·3, \mu = 1002·8, \sigma = 10$.

24. $\sigma = 15, \mu = 50·3$.

25. 28·9, 76·5, 145·3, 163·1, 108·3, 53·9.

26. 0·177, 0·127. 28. (a) 0·0018, (b) 0·9424.

29. Mean 32·093, standard deviation 0·0868, new mean 32·143, 14·2 per cent.

30. 3 per cent, 17 per cent, 37 per cent, 31 per cent, 10 per cent, 1 per cent, above 0 per cent, below 0 per cent, to the nearest unit.

Exercises 9a

1. $z = 1·8$: (a) do *not* reject the null hypothesis at the 5 or 1 per cent levels, (b) do *not* reject the null hypothesis at the 1 per cent level but do so at the 5 per cent level.

2. $z = 2·5$; reject the null hypothesis at the 1 per cent level.

3. Reject the idea that the coin is biased at 5 per cent level but *not* at 1 per cent level.

4. The results are not significant at the 5 per cent level.

Exercises 9b

1. The sample supports the idea of change.

2. No.

3. $z = 3·18$, the claim is rejected at the 1 per cent level of significance.

Exercises 9c

1. (a) reject the null hypothesis, (b) do *not* reject the null hypothesis.

2. The result is not significant.

Exercises 9d

1. The treatment has a significant effect.

Exercises 9e
1. 24.12–35·88.   2. 28·82–31·18.
3. 157·34–166·66.

EXERCISES 9

1. (*a*) 2·68, (*b*) 1·78, (*c*) 1·78, (*d*) 2·18.
2. (*a*) 2·15, (*b*) 3·25, (*c*) 1·70, (*d*) 2·55.
3. (*a*) 1·72, (*b*) 2·81, (*c*) 2·76.
4. (*a*) 1·29–1·75, (*b*) 1·35–1·69.
5. $t = 0·789$, not significant.   6. No evidence
7. $t = 2·6$, not significant.
8. $\bar{x} = 6·76$, $s = 4·86$.
9. No significant improvement.   10. $n \geqslant 2398$.
11. $t = 2·64$, evidence supports claim.
12. $z = 2·5$, results significant at the 1 per cent level.
13. $z = 3·8$, the results indicate a difference in strength.
14. No.
15. 139·25–146·45.   16. Zad is a very inferior product.
17. (*a*) reject, (*b*) reject.   18. Pop music is better.
19. 2310, 2258·4–2361·6.   20. Significant at the 1 per cent level.
21. No evidence, evidence at the 1 per cent level.
22. Significant at the 5 per cent level but not at the 1 per cent level.
23. Mean 0·4154, 0·2062 significant difference.
24. Significant difference at 1 per cent level.
26. Significant difference.

Exercises 10a
1. Mean 1·50309–1·50710, 1·50190–1·50829.
   Range 1·98–9·68, 0·86–12·63.

Exercises 10b
1. 0·012–0·078.   2. 2 and 3.

Exercises 10c
1. 1·78–6·54, 0·41–7·91.

EXERCISES 10

1. Mean 5·095–5·997, 4·765–6·327.
   Range 0·133–1·607, 0·029–2·215.

2. Mean 2·782–6·59, 1·684–7·68.
   Range 3·197–11·537, 1·807–14·734.
3. Mean 1·277–1·299, 1·27–1·306.
   Range 0·0111–0·0543, 0·0048–0·0708.
4. Inner 3, outer 4.
   allowable width 1·2786–1·3004.
5. Inner 4, outer 8, 25 per cent.   6. 0·2231.
7. Mean 19·998–20·012.
   Range 0·00192–0·02832.

Exercises 11a
1. (*a*) 23·21, (*b*) 6·84, (*c*) 57·49, (*d*) 51·11, (*e*) 62·77, (*f*) 158·17, (*g*) 314·35.
2. (*a*) 0·9, (*b*) 0·657, (*c*) 0·1, (*d*) 0·0067.

Exercises 11b
1. $\chi^2 = 60·82$, not significant.   2. No.

Exercises 11c
1. $\chi^2 = 74·9$, reject at the 1 per cent level.

Exercises 11d
1. 2·84–6·75.

Exercises 11e
1. (*a*) $\chi^2 = 9·34$ which is just significant at the 1 per cent level, therefore reject $p = \dfrac{1}{6}$, (*b*) $\chi^2 = 4·85$ which is *not* significant at the 1 per cent level, therefore do not reject the hypothesis that the data comes from a binomial distribution.

EXERCISES 11

1. (*a*) 21·03, (*b*) 18·48, (*c*) 6·84, (*d*) 23·3, (*e*) 99·65, (*f*) 152·4, (*g*) 127·9.
2. (*a*) 0·8, (*b*) 0·9, (*c*) 0·69, (*d*) 0·1.
3. $\chi^2 = 4·25$, we cannot reject the null hypothesis.
4. 0·962–39·1.
5. $\chi^2 = 3·56$, we cannot reject the hypothesis that the die is fair.
6. $\chi^2 = 5·72$, we cannot reject the hypothesis that the numbers are a random set.
7. $\chi^2 = 3·864$, just significant at the 5 per cent level.
9. $\chi^2 = 11·55$; reject null hypothesis at the 1 per cent level.
10. $N = 519$.

11. $\chi^2 = 1.67$, we cannot reject that the results are consistent.
12. Not significant at the 5 per cent level.
13. $\chi^2 = 2.91$, not significant at the 5 per cent level.
15. $\chi^2 = 1.88$, we cannot reject that short-sightedness is independent of sex.
16. $\chi^2 = 5.71$, result confirms the hypothesis.
17. $\chi^2 = 0.14$, results do *not* indicate a difference.
18. $\chi^2 < 1$, we cannot reject the hypothesis that the results follow a Poisson distribution.
19. $\chi^2 = 7.94$, the figures do not indicate a significant difference.
20. $\chi^2 = 16.8$, we cannot reject the hypothesis that hair colour and place of abode are not related.
21. Cannot reject that they follow a Poisson distribution.
22. $\chi^2 = 30.78$, the performances are related.
23. (a) $\chi^2 = 18.72$, cannot reject that grades and colleges are not related,
   (b) $\chi^2 = 12.74$, A and B turn out more honours graduates,
   (c) $\chi^2 < 1.0$, cannot reject that they are in the ratio $1:3:3:1$,
   (d) $\chi^2 = 1.81$, cannot reject that there is no preference.
24. $\chi^2 = 7.33$, reject hypothesis.   25. No.
26. $\chi^2 = 3$ when treated as a contingency table, and the null hypothesis of no effect cannot be rejected. He is wrong, with a null hypothesis of no effect, he has two groups which are the same and they should be combined to find the expected number of multiple births.
27. Not significant at 5 per cent level.
28. Not significant at 5 per cent level (take expected values to be > 3).
29. Just *not* significant at the 5 per cent level.

EXERCISES 12

1. (a) 6.06, (b) 4.20, (c) 2.51, (d) 0.162, (e) 0.212, (f) 0.248
3. $F = 1.59$, we cannot reject that the populations have the same variance. Assumptions: independent random samples, normally distributed populations.
4. $F = 1.87$, we cannot reject that the variances are the same.
5. 0.353–12.81.
6. By graph 0.04, by linear interpolation 0.0426.
7. 0.156–10.8.

Exercises 13a

1. $r_s = -0.2517$, $r_k = -0.182$, both give non-significant results.

EXERCISES 13

2. $y = 1.2x + 3$.
3. $y = 0.606x + 29.88$, $x = 0.525y + 49.07$.
4. $r = 2.287$, $R = -0.0784$.   6. $\pm r = 0.632, 0.444, 0.279, 0.197$.
7. $r = 0.849$, $\dfrac{s_x}{s_y} = 0.707$.
8. (a) $t = 1.18$ not significant at the 5 per cent level.
   (b) $z = 1.51$ not significant at the 5 per cent level.
9. $r = \pm 0.424$.
10. $y = 0.429x - 0.706$, $x = 1.444y + 1.678$.
11. $r = 0.825$, reject null hypothesis.
12. $r_s = 0.891$, $r_k = 0.733$.
13. $r_s = -0.421$, caanot reject null hypothesis.
14. $r_s = 0.5$.
15. In choosing the second set at random there are several random arrangements (including the one obtained in the first set) which could occur (each with the same probability) and give a significant correlation.
16. Not significant at the 5 per cent level.
17. The rankings are not correlated.
18. £230.   19. The gradient $\simeq 2$.
20. 33.7–51.3.   21. $r = 0.783$.

EXERCISES 14

1. Mean = 85, standard deviation 5.03.
3. Mean 1.6, standard deviation 0.5, $p = 0.212$.
4. Mean 125.4, standard deviation 1.6.
5. 0.081.
6. 0.527.   7. 7.71–9.09.
9. The average of 120 is probably not corrected since account is not taken of any correlation between hours worked and wages per hour.
10. Mean 7, variance 17.5.
11. Mean 3 min 49 sec, standard deviation 2.015 sec, Pr ($<$ 3 min 45 sec) = 0.0235.

Exercises 15a

1. 7s. 6½d.   2. 50·9.
3. £705.
4. September £12·1 per ton; October £12·6 per ton. Although the average price of coal at each depot falls in October, the overall price increases. This is due to the relatively large percentage change in the quantity of 'dear' coal sold, compared with the smaller percentage changes in the other types.
5. 3·118.

Exercises 15b

1.

1905	1906	1907	1908	1909	1910	1911	1912	1913	1914
90·1	93·0	108·8	109·0	96·4	100	97·5	108·4	118·9	116·5

2.

1950	1951	1952	1953	1954	1955	1956	1957	1958
90	96·7	99·2	100	100	100	101·7	105·0	107·5

3. 127·69.

Exercises 15c

1. Crude death rate 18·83; standardized death rate 10·26.
2. Crude unemployment rate 350; standardized unemployment rate 58·5.

Exercises 15d

2. The 12 month moving averages are 918, 850, 772, 714, 665, 621, 583, 540, 497, 447, 394, 344, 299, 258, 228, 207, 188, 171, 159, 148, 140, 134, 128, 123, 119.
3. The 3 day moving averages are 24·3, 21·0, 20·3, 19·0, 13·3, 9·0, 7·0, 7·7, 5·0; 5 day moving averages 23·0, 19·8, 16·0, 13·8, 11·4, 8·4, 5·4.
4. 67, −98, −127, 160.
5. The length of the seasonal variation is 5 years. The yearly seasonal variations are 6, 205, 156, −176, −194. The residual variations from 1942 onwards are respectively 1, 3, −1, 0, 1, −2, 3, 2, 0, −3, 5, 2, −2, −1, 3, −2.

EXERCISES 15

1. 2·63.   2. 62·39 d./gal.
3. The 4 quarter moving averages are as follows 10·0, 10·6, 11·3, 12·1, 12·3, 13·4, 13·2, 12·9, 13·2, 12·8, 13·7, 15·0, 16·3, 18·2, 19·5, 20·7, 22·2.

4.

	1932	1933
1st quarter	31·3	32·8
2nd quarter	26·3	20·8
3rd quarter	27·6	19·3
4th quarter	30·6	22·0

5. The 12 month moving averages are 14·2, 14·0, 14·0, 14·1, 14·1, 14·1, 14·2, 14·4, 14·5, 14·6, 14·6, 14·8, 15·0, 15·3, 15·4, 15·3, 15·4, 15·6, 15·6, 15·7, 15·8, 15·8, 15·8, 15·7.

6. Standardized death rate 13·98.

7. Quarterly variations 4·2 per cent, −3·2 per cent, −2·9 per cent, 1·7 per cent.

8. (a) 2·917, (b) 1·643.   9. 105·7, 96·5.

11. (a) 20, (b) 46, (c) 126, formula S − A + 1.

12. 27s. 8½d. Yes. If the quantities sold at the cheaper prices increased and the quantities sold at the dearer prices fell.

13. Corrected seasonal variations 4·51 per cent, −3·17 per cent, −2·23 per cent, 1·02 per cent; residual variations 0·06, 0·40, −0·76, 0·52, −0·12, −0·55, 0·99, −0·38.

14. The seasonal variations using a 5 year moving average are

$$+6·4$$
$$-2·6$$
$$-14·2$$
$$+7·8$$
$$0$$
$$-6·4$$
$$+5·6$$
$$+8·8$$
$$-9·8$$
$$-3·0$$

The supposed seasonal variations for corresponding years vary so much that it is not likely that there is a 5 year seasonal trend.

INDEX

289